Springer Tracts in Advanced Robotics

Volume 131

More information about this series at http://www.springer.com/series/5208

Frank Bonnet · Francesco Mondada

Shoaling with Fish: Using Miniature Robotic Agents to Close the Interaction Loop with Groups of Zebrafish *Danio rerio*

Springer

Frank Bonnet
Biorobotics Laboratory, Ecole
Polytechnique Fédérale de Lausanne (EPFL)
Lausanne, Vaud, Switzerland

Francesco Mondada
Biorobotics Laboratory, Ecole
Polytechnique Fédérale de Lausanne (EPFL)
Lausanne, Vaud, Switzerland

ISSN 1610-7438 ISSN 1610-742X (electronic)
Springer Tracts in Advanced Robotics
ISBN 978-3-030-16783-7 ISBN 978-3-030-16781-3 (eBook)
https://doi.org/10.1007/978-3-030-16781-3

Library of Congress Control Number: 2019936278

This Springer imprint is published by the registered company Springer Nature Switzerland AG
The registered company address is: Gewerbestrasse 11, 6330 Cham, Switzerland

Many men go fishing all of their lives without knowing that it is not fish they are after.

—Henry David Thoreau

Foreword

Robotics is undergoing a major transformation in scope and dimension. From a largely dominant industrial focus, robotics is rapidly expanding into human environments and vigorously engaging in its new challenges. Interacting with, assisting, serving, and exploring with humans, the emerging robots will increasingly touch people and their lives.

Beyond its impact on physical robots, the body of knowledge robotics has produced is revealing a much wider range of applications reaching across diverse research areas and scientific disciplines, such as: biomechanics, haptics, neurosciences, virtual simulation, animation, surgery, and sensor networks among others. In return, the challenges of the new emerging areas are proving an abundant source of stimulation and insights for the field of robotics. It is indeed at the intersection of disciplines that the most striking advances happen.

The *Springer Tracts in Advanced Robotics (STAR)* is devoted to bringing to the research community the latest advances in the robotics field on the basis of their significance and quality. Through a wide and timely dissemination of critical research developments in robotics, our objective with this series is to promote more exchanges and collaborations among the researchers in the community and contribute to further advancements in this rapidly growing field.

The monograph by Frank Bonnet and Francesco Mondada is based on the first author's doctoral thesis. It effectively brings together concepts from robotics and biomimetics. Novel tools are introduced that can integrate shoals of zebrafish and characterise their collective behaviours. Bio-inspired controllers are designed for the robots to mimic fish locomotion underwater.

Rich of examples developed by means of extensive experimentation on mixed groups of fish and robots, this volume was the co-winner of the 2018 Georges Giralt Ph.D. Award for the best doctoral thesis in Europe. A very fine addition to the STAR series!

Naples, Italy
February 2019

Bruno Siciliano
STAR Editor

Acknowledgements

This work was part of a multidisciplinary project involving many people from different fields for several years, and its achievement would not have been possible without their contributions.

First and foremost, I would like to thank my supervisor Francesco Mondada. Francesco taught me how to manage such a multidisciplinary project with many partners from different backgrounds. He offered me the freedom in my work, while orienting me during the harder times. He gave me the opportunity to teach at the university level during the time of this thesis, and provided me with a great research environment, with dynamic people and enriching discussions, which improved a lot the quality of this work.

This thesis would also not have been possible without the work, help and friendship of Alexey Gribovskiy. Alexey is the person who first hired me for a semester project that involved the first design of one of the robots that is presented in this thesis, and he is the one who first convinced me of doing a Ph.D. He also contributed to the project proposal that financed my thesis, and, finally, advised and contributed largely to this work. Alexey was always a great motivator and his expertise in this field was very precious.

Then, I would like to thank the members of my jury: Alcherio Martinoli, Auke Ijspeert, Jean-Louis Deneubourg and José Halloy for taking the time and effort to read my thesis, travel to Lausanne to assist to my thesis defense, and for their valuable comments and suggestions which helped me to improve the final version of this manuscript. I thank also José for his crazy ideas that made this project possible and for his constructive remarks during the project.

I am grateful to the EPFL and the doctoral school program for providing a great studying and working environment, and to the European Commission whose funding made this research possible.

Among the different colleagues that worked with me on this project, I would like to thank first my colleagues in the Mobots group. I want to give a special thanks to Daniel Burnier and Norbert Crot, who shared their experience with me and helped me a lot to improve my work on many aspects during the project. Dany and Norbert also invested a lot of energy to design various parts of the robots and experimental setup, and for that I am very grateful. Dany and Norbert also brought a lot of positive atmosphere in the open space, which made every work day a pleasure to be there. I also would like to thank Philippe Rétornaz for his precious advices and coding capacities that accelerated the development of the project in its early phase. Thanks go to Christophe Barraud who replaced me during the first phase of the project and realized progress on the robotic design, and to Marcelo Elias de Oliveira for his contribution on the software side. I am grateful to Manon, Léa and Maria for their designer expertise that improved the presentation of this work. I thank also the rest of the group with which it was a pleasure to work with: Anne, Claudia, Vincent, Fanny, Florian, Stefan, Gordana, Morgane, Yui, Olguta, Hadi, Mariza, Michael, Ehsan, Stéphane and Martin.

I would like to express my gratitude to all my colleagues in the consortium of the ASSISIbf project. I would like to especially thank our colleagues from the University Paris Diderot, in particular, Bertrand and Leo for their work on the biological and coding aspects respectively, but also Axel and Yohann for their patience in performing the experiments with the robotic infrastructure. I thank also Bertrand who took the time to read and make precious comments on the final version of this manuscript. I am also grateful to Thomas Schmickl and his team at the University of Graz, for coordinating the project, and for bringing positive and constructive remarks on my work during the project meetings.

Several students and interns also contributed to this work. I would like to thank especially Yuta for his work on the design of the RiBot, and Matthieu for his design of the biased circular arena and for his contribution in the production of the FishBots. I also thank Cynthia, Samuel, Patrick, Shujie and Laila for their hard work that contributed to this thesis.

I also want to thank Marcel Gyger and Xavier Warot for helping me getting the authorization to perform the experiments with zebrafish at the EPFL. I also thank the cantonal veterinary for authorizing me to perform these experiments that were necessary to qualify the robotic system.

This work would not have been possible without the support of my family and friends. I am deeply grateful to my parents, Michéle and Jean-Luc, for their support and encouragements during my life, and especially throughout my studies. Thanks also to my sister Sandra for helping me sorting my Legos during my childhood, and to my grandparents, Janine and Robert for their continuous support and lifestyle example. I have some gratitude also to my family-in-law, especially to Anna and Sergei, for always welcoming me well during my trips to Moscow and for interesting conversations.

I also have been able to count on many friends for unforgettable memories and making my mind escape from my thesis. I would like to give a special thanks to Philipp with whom we faced the challenge of the thesis together during courses and coffee breaks in Sat. I thank also Alex, Thy, Marc and Orane, and, last but not least, the Champions' team: Hannah, Ludwig, Tristan and Julien.

I would like to thank deeply my wife, Jenia, for her patience, support and love during this work, and for making my life a paradise every day. Jenia is the best thing that ever happened to me, and it is not a hazard that I could achieve and obtain some great results during the time of the thesis, and I am endlessly grateful to Jenia for that.

Finally, I would like to thank all the zebrafish that participated in my experiments and with which we have shared some difficult moments. Making experiments with animals taught me also lessons on how to respect and admire more the animal world that surround us, and that, most of the time, the capacities and intelligences of these organisms are neglected by us. I hope that these fish also enjoyed their time in our laboratory and the company of our robotic fish.

Lausanne, Switzerland
March 2017

Frank Bonnet

About This Book

Robotic animals are nowadays developed for various types of research, such as bio-inspired robotics, biomimetics and animal behavior studies. The miniaturization of technologies and the increase in performance of embedded systems allowed engineers to develop more powerful, sophisticated and miniature devices. The case of robotic fish is a typical example of such challenging design: the fish locomotion and body movements are difficult to reproduce and the device has to move autonomously underwater. More specifically, in the case of collective animal behavior research, the robotic device has to interact with animals by generating and exploiting signals relevant for social behavior. Once perceived by the animal society as conspecific, these robots can become powerful tools to study the animal behaviors, as they can at the same time monitor the changes in behavior and influence the collective choices of the animal society.

In this work, we present novel robotized tools that can integrate shoals of fish in order to study their collective behaviors. This robotic platform is composed of two subsystems: a miniature wheeled mobile robot that can achieve dynamic movements and multi-robot long-duration experiments, and a robotic fish lure that is able to beat its tail to generate fish-like body movements. The two subsystems are coupled with magnets which allow the wheeled mobile robot to steer the robotic fish lure so that it reaches very high speeds and accelerations while achieving shoaling. An experimental setup to conduct studies on mixed societies of artificial and living fish was designed to facilitate the experiments for biologists. A software framework was also implemented to control the robots in a closed-loop using data extracted from visual tracking that retrieved the position of the robots and the fish. We selected the zebrafish *Danio rerio* as a model to perform experiments to qualify our system. We used the current state of the art on the zebrafish social behavior to define the specifications of the robots, and we performed stimuli analysis to improve their developments. Bio-inspired controllers were designed based on data extracted from experiments with zebrafish for the robots to mimic the zebrafish locomotion underwater.

Experiments involving a robot with a shoal of fish in a constrained environment showed that the locomotion of the robot was one of the main factors to affect the collective behavior of zebrafish. We have also shown that the body movements and the biomimetic appearance of the lure could increase its acceptance by fish. Finally, an experiment involving a mixed society of fish and robots qualified the robotic system to be integrated among a zebrafish shoal and to be able to influence the collective decisions of the fish. These results are very promising for the field of fish-robot interaction studies, as we showed the effect of the robots in long-duration experiments and repetitively, with the same order of response from the animals.

Keywords Animal-robot interaction • Collective behavior • Mobile robotics • Biomimetic robots • Underwater robotics • Visual tracking • Multi-agent system • Zebrafish

Contents

List of Figures

List of Tables

Chapter 1
Introduction

The goal of this thesis is to contribute to the field of animal-robot interaction. Using zebrafish as a model of a social animal that is capable of collective decision-making, and by developing miniature robotic tools to integrate shoals of zebrafish, we aim to create mixed societies of living and artificial agents that will provide scientists with a new framework for animal ethological studies. This chapter presents the motivation behind this area of research, followed by a description of the methodology applied. The contributions of the thesis are then presented, followed by a detailed description of the thesis' organization.

1.1 Motivation and Challenges

1.1.1 The Study of Collective Animal Behaviors

Have you ever wondered, while walking along a lake shore, or simply when peering into an aquarium, how fish communicate with each other and what mechanisms cause them to aggregate and form sometimes extremely complex structures? In general, what is it that leads social animals to interact, and how do they take collective decisions? Of course, fish are not the only animal that can create such complex coordinated systems; it is a trait shared by most other living organisms. These impressive phenomena have raised many questions, which are yet to be answered.

The aim of collective animal behavior studies is to examine how local interactions between individuals that are parts of a group can combine to produce global-level outcomes for the animal society (Camazine et al. 2001). Indeed, by aggregating and taking collective decisions, animals can sometimes even solve problems that an individual could not solve alone. Ants are a notable example of this (Detrain and Deneubourg 2006; Czaczkes and Ratnieks 2013). The study and modeling of the collective mechanisms inside animal societies, such as the transfer of information

F. Bonnet and F. Mondada, *Shoaling with Fish: Using Miniature Robotic Agents to Close the Interaction Loop with Groups of Zebrafish Danio rerio*, Springer Tracts in Advanced Robotics 131, https://doi.org/10.1007/978-3-030-16781-3_1

and exploration strategies, has even inspired an entire field of computer science (Bonabeau et al. 1999), with new types of algorithms created to optimize complex tasks within our own societies (Dorigo et al. 2006).

1.1.2 Fish as a Model of a Gregarious Animal

The collective behavior of fish, in particular, has piqued the interest of scientists. Many fish species are indeed social and can form groups that range in number from several to hundreds of individuals, sometimes achieving complex formations. Some small species of fish are also ideal candidates for the study of collective animal behavior, as they can easily be bred and observed under laboratory conditions. When placed in an appropriate tank, fish will exhibit shoaling or schooling behavior; therefore, by using appropriate tracking tools and ensuring the right environment, it is possible to observe their behavior for long periods.

When aggregating, fish can form in general two types of structure: the shoal or the school. The shoal is a simple grouping of fish, while the school is composed of individuals that swim in a synchronized and polarized way (Pitcher and Parrish 1993). As with other animals, there are several reasons for fish to band together (Wright et al. 2006). First, shoaling offers some level of protection against predators. Indeed, a group composed of several individuals can observe the external environment more widely, and when an individual senses the presence of a predator, it can quickly emit a signal to warn its conspecifics. This phenomenon is also known as the *many eyes hypothesis* (Olson et al. 2015). Another advantage is that it is harder for a predator to attack a shoal than a solitary animal, due to the confusion induced by the combined motion of many individuals. Shoaling also enhances the group's own hunting process, as it is more efficient to hunt within a group. Finally, it has recently been shown that shoaling can reduce the amount of energy expended by individuals when moving; this is also the reason why some birds fly in a V formation (Ashraf et al. 2016).

Scientists have identified certain communication channels used by fish, such as sound, body gesture signals, chemical cues, and color changes (Ioannou 2016). It is also believed that the fish exchange information and maintain their group formation using passive cues, based on the movements of their neighbors, that are sent by the vision organs and the lateral line, an organ used to sense water pressure (Dykgraaf 1933). However, compared to other animals, it is still not entirely clear which communication channels are involved in the triggering of social behavior in fish, as it has also been shown that fish can form mixed groups composed of several species, for instance when sharing the same food diet (Pitcher and Parrish 1993). In addition, the mechanism of collective decision-making and leadership in fish still requires more in-depth investigation (Ward et al. 2013).

When it comes to studying the collective behavior of fish, one should select a good candidate from the thousands of gregarious fish species. One fish that is commonly used as a model for behavioral studies, as well as in other fields, is the zebrafish *Danio rerio*. One advantage of this fish is the ease of breeding under laboratory conditions.

It is a freshwater fish of small size (~40 mm standard length). As it is a species that is commonly used in research, there is a large amount of prior literature that concerns it. The zebrafish is commonly used in behavioral genetics, as it shows the same types of behavior as mammals, including reward behavior, learning and memory, aggression, anxiety, and sleep. In addition, many protocols have been developed to study the zebrafish individual and social behaviors (Norton and Bally-Cuif 2010). Finally, compared with fish species that swim in schools, such as tuna or pilchard, the zebrafish swims most of the time in shoal. This makes it a good candidate to studying the collective behavior of a group, as "schools" can almost be considered as individuals who share a common behavior.

1.1.3 Methods of Studying the Collective Behavior of Animals

The scientific literature provides us with many examples of various mock-ups that are used to study social communication, relationships, and structures in animal groups. The first experiments of the twentieth century to explore these mechanisms were designed to test just one specific behavior, such as that described by Tinbergen (1951), where it was shown that animal communication can be based on simple signals and that social interaction with animals can be established by creating specifically designed artifacts that generate and exploit only a part of the signals that are relevant to social behavior.

In order to accomplish this, it is first necessary to acquire an adequate knowledge of the various animal organs' sensory capacities. Then, one should be able to set up an experiment to determine whether the perceived stimuli generate a reaction. In his work *Study of Instinct*, Tinbergen illustrates the example of the stickleback, which can be provoked using a fake fish (Fig. 2.1), regardless of the size and shape of the fake fish, as soon as the stickleback saw that it had a red belly, it was attacked. This is because, when two stickleback fight, both their belly and throat turn red, which generates a reaction in the opponent. This essentially proves that the stickleback reacts to red and appears to neglect most of the other characteristics, even if it is known that the stickleback has very keen eyesight and is able to distinguish many other details (Tinbergen 1951).

To stimulate the behavior of the animal, researchers usually use simple apparatus to separate the various stimuli perceived by the animals, as per Tinbergen's approach, while using the most recent technology. For instance, Bass and Gerlai (2008) used stimulus fish on screens and measured the zebrafish's response to show how the fish perceive predators differently from harmless fish. Meanwhile, Abaid et al. (2012b) used images animated according to mathematical models of animal grouping and assessed the zebrafish's response. Finally, Fleisch and Neuhauss (2006) showed various apparatus used to characterize the visual response of zebrafish.

Despite the fact that these methods can indeed trigger some specific behaviors in the animal, they cannot support complex interactions with the animals. The idea of using agents to infiltrate inside the animal's social system could be beneficial in terms of joining in with their activities, and monitoring their behavior while also being able to adapt to the animal under studies, in order to learn how the animals exchange information that generates changes in behavior in various contexts. In recent years, the relevant technology has become more advanced and affordable, and robotic devices have emerged that are capable of sending cues to the animals, sensing their responses, and adapting to their behaviors. This dissertation will thus demonstrate how specifically designed robots can be used for such purposes with zebrafish.

Among the recently developed tools to study the collective behavior of fish, tracking frameworks have also become highly accurate. Improvements to cameras, image processing, probabilistic approaches, and the use of three-dimensional tracking are some of the latest advancements that have allowed researchers to track a great number of individuals within a fish shoal. Such observations enable the automatic characterization of the fish's orientation and speed over time, as well as their inter-individual distances (Delcourt et al. 2013). For instance, Pérez-Escudero et al. (2014) designed tools that allow the tracking of individuals within groups of fish. Even if such software cannot yet perform individual tracking in real time, it allows researchers to characterize the behavior of individuals inside shoals, especially when coupled with apparatus that is able to send specific stimuli. Therefore, the current technology permits the application of a closed-loop system that can adapt the generation of stimuli in function of fish behavior. With this in mind, we will also present a new framework that allows closed-loop control of robots that interact with animals and can adapt depending on changes in the animal's behavior.

1.1.4 A Benefit for Farming and Ecology

Social animals can either be considered a necessity for humans, such as in the case of a cattle herd, or a threat, as in the case of a medusa swarm. In both cases, the control and monitoring of these groups may improve human welfare.

In the farming industry, the breeding of animals is crucial in terms of benefiting the companies, as well as for the well-being of the animals (Munack 2002). The use of agents that are capable of autonomously interacting with the animals could allow the detection of potential risks to the health of the animal, and with the option of modifying the animal's behavior to resolve this issue, especially in a human society that is moving continuously towards data logging and the Internet of Things, in which all objects are connected (Madakam et al. 2015).

If we look at the case of wild animals, fish in particular have certain sensory properties that could benefit the entire ecosystem. Indeed, it has been shown that certain species of fish can detect pollutants in water (Travis 2013). Nowadays, animals suffer greatly from pollutants generated by humans, especially in the agricultural industry, such as shown by Magalhães et al. (2007, 2012). Indeed, large corporations continue to produce many different products (herbicides, fungicides, or insecticides) that are used by farmers to keep their fields in good condition, despite the fact that this contaminates the environment that is essential for many animal species. For example, one of the major causes of bee extinction is pollutants that bees bring back to the hive (Goulson et al. 2015). Another is an increase in chemical components, stemming from the products used in farming, in rivers and lakes, which causes the death of many types of fish (Zeitoun and Mehana 2014). Fish could be used as ecological biomarkers for the detection of pollutants (Travis 2013). Indeed, scientists have observed certain changes in fish behavior in the presence of pollutants, including in zebrafish (Giattina and Garton 1983). Accordingly, by having agents inserted among animal societies, one could monitor and track changes of behavior in the animals. Through this process, the control of the fish shoal could also be beneficial for the fish in case of a situation in which the animals could not find a solution to avoid a polluted area on their owns.

1.2 The Field of Animal-Robot Interaction

1.2.1 The Use of Robots to Study Collective Animal Behaviors

As already mentioned, in recent years, thanks to the development of technologies, especially in the research field of robotics, sophisticated devices that able to interact with animals have appeared. Robots do indeed offers several advantages in the study of the animal behavior (Garnier 2011): they can be built and modulated as required, mainly based on the new technologies; they are physical entities, and therefore can have real physical interactions with the animals. In addition, during the development of these robots, scientists must find the exact communication channels that the animals use and that, for instance, trigger social behaviors among conspecifics, and which brings benefits for our general knowledge of animal communication. Finally, robots are appropriate tools for testing models of behavior, as, in comparison with simulators, they can reproduce the same dynamics and actions as the animals in the environment (Krause et al. 2011).

In terms of ethics, mobile robots—robots that are capable of locomotion—are a good alternative for studying the animals in their environment without being too intrusive. In fact, we would probably be able to collect certain interesting information or even control animals' actions by plugging electrodes into animals' brains to measure their activity, or inserting an embedded system inside the animal (Sato and Maharbiz 2010). Sometimes, there are no alternatives to the use of such intrusive devices in the performance studies into animals, however, we think that the integrity of the animal should be preserved, and if systems that are able to study the animals' behavior exist to offer an alternative to intrusive systems, their use should be encouraged over the other option.

Given the above, the use of mobile robotic devices to study the behavior of animals is increasing, both to study the animals in their natural environments (Le Maho et al. 2014) and in laboratory research. While the behavior of animals can be observed by the robots, depending on their design, can also be influenced by them.

In Vaughan et al. (2000), a robotic sheepdog that was able to gather a flock of ducks was introduced (Fig. 1.1a). This was the first example of a robotic device exploiting and controlling the animal behavior in a closed-loop, thanks to the Shepherd dog's effect on the ducks. In contrast to the research to be presented in this dissertation, the robot was not perceived as a conspecific by the animals, but as an external agent that was able to control the behavior of the animals.

The first project to deal with mixed animal-robot societies involving multiple robots and animals, and where the robots were accepted by the animals as society members, was the European project LEURRE (Halloy et al. 2007). During this project, a mixed society of cockroaches and robots was created, where specially designed autonomous mobile robots were able to interact with cockroaches, and as members of the society, could participate in social decision-making (Fig. 1.1b).

Following the LEURRE project, a project has dealt with mixed groups of robots and chicks (Gribovskiy et al. 2010) (Fig. 1.1c). In this project, it was demonstrated that robots can be successfully socially integrated into the animal group, thanks to the imprinting mechanism, which was confirmed by the following behavior demonstrated by imprinted animals. However, this project did not fully succeed in closing the interaction loop between the animals and the robots, as was achieved in the case of LEURRE. Therefore, no projects to date have successfully created a mixed society of vertebrates with groups of robots interacting in a closed loop with the animals.

Robotic tools are, however, widely used these days in behavioral studies of various species, such as honeybees in Griparic et al. (2015) (Fig. 1.1d) and in Landgraf et al. (2010), rats (Shi et al. 2010), crickets (Kawabata et al. 2013), squirrels (Rundus et al. 2007), Tungara frogs (Taylor et al. 2008), cows (Correll et al. 2008), and most likely many others, with some of them also providing closed-loop control of the robotic agents, which allows the robot to adapt to the animal behavior.

1.2.2 The ASSISIbf Project

The work presented in this dissertation was part of the European FP7 project ASSISIbf (Animal and robot Societies Self-organize and Integrate by Social Interaction with bees and fish) (Schmickl et al. 2013), which is the largest European project of the European Union call FOCAS (Fundamentals of Collective Adaptive Systems) and that received 6 million Euros for a five years period. ASSISIbf has started on February 2013.

The main goal of ASSISIbf is to establish a robotic society that is able to develop communication channels with animal societies (in this case honeybees and zebrafish) on its own using the same methodology that was used in Halloy et al. (2013). These robots will adapt thanks to evolutionary algorithms until they have learned to interact with animals in the desired manner. This innovative technology is aimed at laying new foundations in terms of how humans can interact with animal societies to manage the environment. In parallel, these mixed societies of animals and robots will represent

Fig. 1.1 Robots designed for studies of animal behaviors. **a** The robotic sheepdog and a flock of ducks (Vaughan et al. 2000). **b** The InsBot mobile robot that interacted with a society of cockroaches during the LEURRE project (Halloy et al. 2007). **c** The PoulBot mobile robot designed to interact with chicks (Gribovskiy et al. 2010). **d** A stationary robotic bees composed of different actuators and sensors to interact with honeybees (Griparic et al. 2015)

a novel kind of bio-hybrid system, as the animals will enrich the capabilities of the machines and vice versa.

Six partners are involved in this project: our group at the Ecole Polytechnique Fédérale de Lausanne (EPFL, Switzerland) and a group at the University of Zagreb (Croatia) are building the robotic devices and software tools for interacting with societies of fish and bees (Fig. 2.3d), respectively. Meanwhile, the University Paris Diderot (France) and University of Graz (Austria) are performing the experiments with the living animals: fish and bees respectively. The University of Lisbon (Portugal) is in charge of developing a multi-agent simulation tool and conducting evolutionary computation. Finally, the company Cybertronica (Germany) is in charge of developing some parts of the robotic bee device and modelling of collective complex systems. Most of the work presented in this dissertation was thus made in collaboration with various partners, especially the group from the University Paris Diderot that dealt with the biology of fish and the development of mixed societies of fish and robots.

1.2.3 The Methodology Used in This Work

The study presented in this dissertation is based on the same methodology that was presented in Mondada et al. (2012) and used by Halloy et al. (2007) and Gribovskiy and Mondada (2010). A schematic of the methodology is presented in Fig. 1.2.

The aim of this methodology is to design robots that are capable of interacting with social animals and participating in their social activities by forming a mixed robot-animal society. Thus, the target of this methodology is to close the loop of interaction between the animals and the robots. It may be noted from Fig. 1.2 that the methodology is strongly iterative, thus each novel change in the robot design, to generate, for example, a new stimulus perceived by the animals, should be tested with the animal group to be validated, and the analysis made on the basis of the animal behaviors are used to improve the robot design itself and so on. Therefore, the reader of this dissertation will encounter several of these loops throughout the reading. We will begin the dissertation with a presentation of the robotic design and the entire infrastructure that was built to run the experiments before the analysis of stimuli that were conducted between the robots and animals. However, chronologically, these tests were often overlapping.

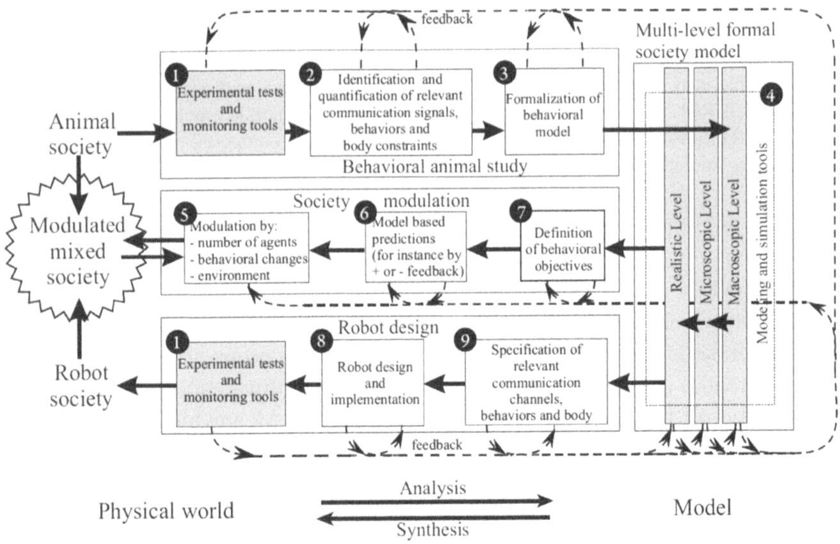

Fig. 1.2 General methodology used for the build-up of an animal-robot mixed society (Mondada et al. 2012). The mixed societies can be defined as "dynamic systems, where animals and artificial agents interact and cooperate to produce shared collective intelligence" (Mondada et al. 2012). This methodology requires the contribution of experts in both fields, biology and robotics, who should work in close synergy as several iterations will be needed during the project to perfect the robotic device

Figure 1.2 indicates the complexity of the various steps of the methodology. In order to summarize it, we simplify the methodology that was applied to this work in four main steps:

- First, we study and model the behavior of the animal. In particular, we examine how the biological organization emerges, as well as the various communication channels used by the animal to transfer information. We also specify the features that are potentially important for the robots' acceptance by the animals. As a result of this step, we obtain a technical specification with which to design the robotic system.
- Then, we design a robotic system following the specifications before verify that it corresponds to them. As we are studying the collective behaviors of the animals, several robots need to be designed in order to build societies of multiple robots, which should reproduce the patterns displayed by the animal society.
- Thereafter, we conduct tests on the acceptance of the robotic devices by the animals. We verify that the robots are triggering the social behaviors of the animals, and not other types of behaviors, such as fear or aggression. This step includes a comparison between a group of only animals with a group composed of robots and animals. The behavior of the individuals should not vary drastically between the two cases, meaning that the robots are acting like them and are integrated.
- Finally, we show that the robots are also capable of influencing the society of animals, just as the members of the society do. While this step is closely connected with the previous one, here we show that the robots can not only act like any member of the animal society, but they can also play the role of leaders among the mixed group.

In this work, we use this methodology for the design of a robotic fish and its control in a closed-loop for interacting with the zebrafish. This involves the electronic and mechanical design of robotic devices, the implementation of software tools to retrieve the fish's behavior and control the robotic devices in a closed-loop, the implementation of behavior-based controllers, and the set-up of experiments to demonstrate the integration of the robots and their ability to influence the collective behaviors of zebrafish.

1.3 Main Contributions of the Thesis

This thesis is highly multi-disciplinary, as it deals with the design of robotic tools that are intended to interact with animals. Its contributions are mainly in the fields of robotics and biology.

1.3.1 Contributions to the Field of Robotics

The first contribution concerns the development of miniature mobile robots that are able to interact with fish underwater. Due to the fact that the robots should be perceived by the fish as conspecifics, and should reproduce the shoaling behavior of zebrafish, they should be of very small size, which push the miniaturization of both electronic and mechanical hardware to its limit. The device should also be waterproof, thus special features should be created to protect the electronics without increasing the size of the system. In terms of maneuverability, due to the fast motion of the fish, the robot should be extremely reactive, and multiple agents should be controlled autonomously for long periods of time. In addition, the devices are specifically designed to be used by our colleagues in the ASSISIbf project, and probably in the continuity of the project, thus highly robust systems, that are able to work for several years, should be built in large quantities.

Another contribution is the development of control methods for multi-robotic agents that should reproduce the movements of fish underwater. Biologists have already provided some high-level models that could reproduce the trajectories of fish in groups; however, they have never been implemented in a closed-loop in a multi-robot platform. In addition, the robotic devices that are already in circulation often do not exactly mimic the locomotion of fish underwater. This work will provide new control methods to make the robotic agents follow fish trajectories, while reproducing the locomotion patterns of fish.

Finally, a software framework that integrates tracking of multiple agents and the control of robots represents a contribution to the field of robotics, which currently has no generic software that can be used for such purposes. By designing our own highly modular and multi-platform software, we hope to help researchers who are concerned with the tracking and control of multi-agent systems.

1.3.2 Contributions to the Field of Biology

We contribute to the field of biology by creating innovative tools for behavioral studies on fish. During the making of these systems, we also enrich the general knowledge on how fish interact together and how these interactions leads to shoaling and collective decision-making.

Throughout this work, we show a methodology that allows to build a system in which robots integrate with groups of zebrafish and interact with them, which has never been achieved before. Although this methodology is mainly based on an existing one that was applied to generate a mixed society of insects and robots (Halloy et al. 2007), this is the first time that it is used to build a mixed society of robots and vertebrates.

Finally, our system also offers perspectives on learning the animals' language and behavior through evolutionary computation. Indeed, biomimetic controllers need to be implemented and parameterized, depending on the behavior of the fish. These parameters could be evolved to precisely identify the change of behavior of the animal, which is allowed by the closed-loop control of the robotic agents.

1.4 Organization of the Thesis

First, in Chap. 2, we describe the current state of the art in terms of research into fish-robot interaction. As our solutions involve the development of miniature underwater robots, we also include a description of the current state of the art in bio-inspired fish-robot design.

Chapter 3 presents the zebrafish, the species that was selected as a model of gregarious fish in our study. We provide a description of the animal's biology, and demonstrate the advantages of using it as a model in various fields of study. We also present the facility that was constructed to breed zebrafish in our laboratory at the EPFL, with a detailed description of the housing aquarium conditions.

Then, Chaps. 4 and 5 present the design of the two robotic devices built for this project to study the behavior of zebrafish. Chapter 4 introduces the FishBot, a fast-wheeled mobile robot designed to steer fish lures so that they move in shoal, while Chap. 5 presents the RiBot, a robotic fish lure that mimics the shape ratio of the zebrafish and can reproduce its tail beating movements. These two chapters are composed of a description of the requirements set for the design, the dimensioning of the various hardware components, an outlining of the different design improvements, a full description of the hardware and firmware, and a validation of the final version under experimental conditions.

In Chap. 6, we present the experimental setup that was designed to perform automated experiments involving mixed societies of fish and robots. We also describe the various arenas that were built to constrain the fish in order to retrieve a clear collective choice response and to increase the acceptance of the robot among the fish society.

The software tools to be used for the mixed society experiments, mostly for controlling the robots in a closed-loop depending on the fish position that was retrieved using image tracking, are described in Chap. 7. The global software architecture is described with the main libraries upon which it depends. In this chapter, we also present all the different navigation algorithms implemented in the various software layers to efficiently control the robots.

In Chap. 8, we illustrate the design of the controllers that mimic the fish movements. There are two levels of controllers that are described: first, the low-level controllers that are implemented at the level of the robots to mimic the locomotion patterns of zebrafish, and second, the high-level controllers that use probabilistic models based on data extracted from observations of the fish in the different types of arenas.

Chapters 9 and 10 demonstrate the results of the stimuli analysis that were performed as part of the design of the robots to qualify them to perform mixed societies experiments. In Chap. 9, we test the impact of several factors that are generated by the robots upon acceptance by the fish, while in Chap. 10, we focus on the types of lures that can be used for this types of research, and more specifically on the need to have lures that are equipped with actuators to mimic the body movements of the fish.

Then, we describe in Chap. 11 an experiment of a mixed society of fish and robots, in which we show how the robots are integrated into the fish society and how they can influence the collective decisions of the fish. A circular corridor is used to reduce the collective decision of the zebrafish to a binary choice, in order to obtain a better measurement of this process. The movements and collective decisions of mixed societies composed of three robots and three zebrafish are compared with groups of six zebrafish in order to measure the impact of the robots on the animal society. The results show that the robots appear to be integrated and are able to monitor the behavior of the fish under different environmental conditions.

Finally, Chap. 12 concludes this dissertation, by summarizing the main contributions of this work to the field of animal-robot interaction. It also describes the perspectives on the future of the research in this field, and how the developed tools could be used for extensive behavior studies, using evolutionary algorithms to obtain an adaptation of the various controllers and behaviors of the robots in order for them to learn the animal's language.

Chapter 2
State of the Art in Fish Behavioral Studies Using Robots and Robotic-Fish Design

2.1 Summary

In this chapter, we present the state of the art in the use of robotized fish lures for fish-robot interaction studies and miniature bio-inspired robotic-fish designs at the time this thesis was written. The last section will describe the position of our work compared to the current state of the art.

2.2 Introduction

In order to study the collective behaviors of the zebrafish, a complete framework needed to be designed, essentially including robots, controllers and software tools. Indeed, since the work of the pioneers of ethology, who began studying animal behavior in the early twentieth century, the required infrastructure has become more complex. We already mentioned, for the case of behavioral studies on fish, the types of lures that were used in Tinbergen (1951) to study, for instance, the communication channels used by the stickleback when fighting (Fig. 2.1). Since then, researchers have started to design robotized lures that are able to generate specific behaviors, sometimes reacting to the animal behavior in a closed-loop control. We will list in Sect. 2.3 the different solutions that can be found in the literature concerning the lures designed for fish-robot interaction studies.

As the aim of this study is to design robotic agents that are perceived by zebrafish as conspecifics, the robots might also need to provide some biomimetics cues, such as activated body parts. We will also mention in Sect. 2.4 some of the existing designs of bio-inspired fish robots.

F. Bonnet and F. Mondada, *Shoaling with Fish: Using Miniature Robotic Agents to Close the Interaction Loop with Groups of Zebrafish Danio rerio*, Springer Tracts in Advanced Robotics 131, https://doi.org/10.1007/978-3-030-16781-3_2

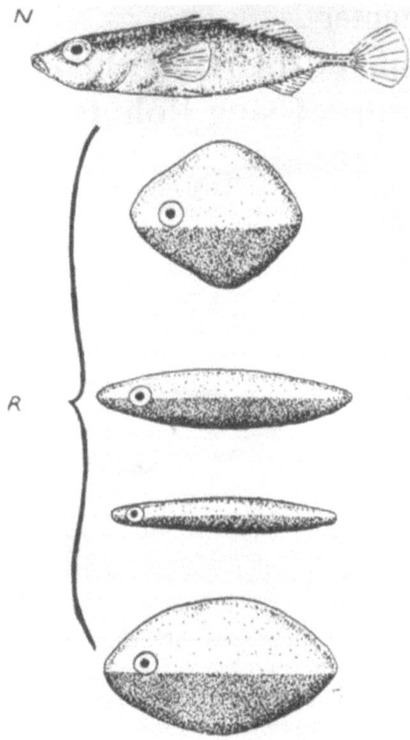

Fig. 2.1 Five stickleback lures, one mimicking the stickleback accurately (N) and four lacking many of the fish characteristics (R) used by Nikolaas Tinbergen in the first half of the 20th century for his research on animal behavior (Tinbergen 1951)

2.3 Robotized Fish Lures for Behavioral Studies

In the past decade, researchers in the field of animal-robot interaction have tried to extend this field to gregarious fish. The challenges of designing robots that have to mimic fish behaviors and visual aspects, while operating underwater, did not stop the scientists, and several groups have thus started to build their own devices. A table summarizing these studies with some key characteristics is presented in Table 2.1, but we will start by first describing the studies in more detail.

In Abaid et al. (2012a), Aureli et al. (2012) and Butail et al. (2014b), the authors observed the zebrafish response to a robotic fish (Fig. 2.2c). This robotic fish, whose preliminary design is described in detail in Kopman and Porfiri (2013), which had the same size ratio as the zebrafish, was attached to a moving device on top of a tank, and its speed, tail beating and coloration could be varied. In Phamduy et al. (2014), a study measured the preference of fertile female bluefin killifish (*Lucania goodei*) for a robotic replica, whose aspect ratio, body size, motion pattern, and color were inspired by an adult male killifish, using a robotic platform specifically designed to simulate the typical courtship behavior observed in male killifish. In Bartolini et al. (2016), a robotic arm was used to move a lure made of a shoal of zebrafish replicas in order to characterize the effect of body size on the social behavior of zebrafish.

Table 2.1 Recent works using biomimetic robots to study the behavior of fish. We classified the studies according to the techniques used to move the lure, the behaviors of the robot, if the robot is control in open or in closed-loop, the shape of the lure and the number of lures tested at the same time in an experiment

Lure motion	Study	Robot's behavior	Robot's control	Shape of lure	Nb of lures
Robotic arm	Phamduy et al. (2014)	Fixed pattern	Open-loop	Biomimetic	1
	Polverino and Porfiri (2013b)	Fixed pattern	Open-loop	Biomimetic	1
	Polverino and Porfiri (2013a)	Fixed pattern	Open-loop	Larger size	1
	Abaid et al. (2012a)	Fixed pattern	Open-loop	Biomimetic	1
	Butail et al. (2014a)	Fixed pattern	Closed-loop	Biomimetic	1
	Ladu et al. (2015a)	Fixed pattern	Open-loop	Biomimetic	1
	Ladu et al. (2015b)	Fixed pattern	Open-loop	Biomimetic	4 fixed
	Polverino et al. (2012)	Fixed pattern	Open-loop	Larger size	1
	Spinello et al. (2013)	Fixed pattern	Open-loop	Larger size	1
	Bartolini et al. (2016)	Fixed pattern	Open-loop	Biomimetic	4 fixed
	Donati et al. (2016)	Fixed pattern	Open-loop	Biomimetic	1
Self-propelled	Abaid et al. (2013)	Fixed pattern	Open-loop	Larger size	1
	Butail et al. (2013)	Fixed pattern	Open-loop	Larger size	1
	Butail et al. (2014b)	Fixed pattern	Closed-loop	Larger size	2
Mobile robot	Faria et al. (2010)	Fixed pattern	Open-loop	Biomimetic	1
	Swain et al. (2012)	Centroid following	Closed-loop	Biomimetic	1
	Landgraf et al. (2013)	Centroid following	Closed-loop	Biomimetic	1
	Landgraf et al. (2016)	Centroid following	Closed-loop	Biomimetic	1

Another study in Donati et al. (2016) used a robotic arm to steer a robotized dummy fish equipped with a moving tail that was able to generate electrical stimuli to attract shoals of weakly electric fish *Mormyrus rume* (Fig. 2.2f).

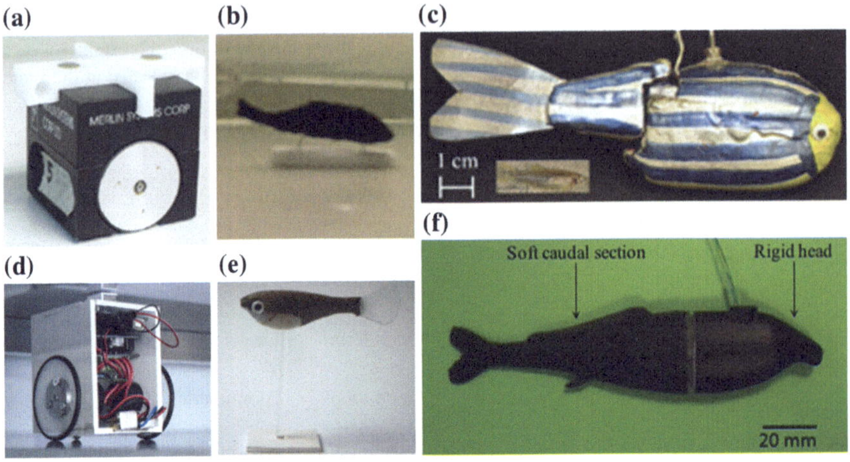

Fig. 2.2 Robots designed for behavioral studies of small species of fish. **a** Commercialized mobile robot *MiaBot* that can steer fish lures **b** inside a tank to study the behavior of golden shiners (Swain et al. 2012). **c** A robotic fish lure that can beat its tail to mimic the body movements of zebrafish. The lure is attached to a robotic arm that can move it underwater (Polverino et al. 2012). **d** *Robofish* mobile robot specifically designed to steer a lure **e** inside a tank in order to study the behavior of guppies (Landgraf et al. 2013). **f** An actuated robotic fish able to emit electric signals for social interaction with weakly electric fish *Mormyrus rume* (Donati et al. 2016)

In Abaid et al. (2013), Butail et al. (2013) and Butail et al. (2014b), the same robot presented in Kopman and Porfiri (2013) was used, but this time self-propelled. This solution allows more robots to be moving in shoal, although the robots do not reproduce the same swimming dynamic as the zebrafish.

In Faria et al. (2010), the authors used a two-dimensional moving platform underneath a tank to transmit the two-dimensional motions to a lure inside the tank. This allowed reproduction of the fish trajectories and locomotion patterns; however, this type of platform did not allow multi-robot experiments.

In Marras and Porfiri (2012), Landgraf et al. (2013), Landgraf et al. (2016), Swain et al. (2012) and Worm et al. (2014), a passive lure attached to a support was moved using a mobile robot below the aquarium and controlled using tracking software. This solution allows for multi-robot experiments, with individual robots able to mimic the fish trajectories. If we look more closely into the mobile robot design, in Landgraf et al. (2013), the mobile robot is composed of an Arduino board, a LiPo battery, and two DC motors, and was developed to move a replica fish through magnetic coupling in order to be accepted by a school of guppies (Fig. 2.2a, b). Another similar system for moving a lure inside an aquarium is presented in Swain et al. (2012), who used an existing mobile robot, the *MiaBot* (Merlin Systems Corp, UK), to move the lure inside the aquarium in order to analyze the interactions between predator and prey in small species of fish (Fig. 2.2d–e).

While these different studies demonstrated the potential to develop artificial devices that are able to interact with fish, there is no solution involving an active robotized-lure with a size close to the size of the fish that can reproduce the pattern motion of the fish and autonomously move in an aquarium to integrate a fish shoal. The fact that in this work we target the creation of a mixed society of fish and robots, in which the robots interact with the fish in a closed-loop implies the design of specifically dedicated robotic devices and software tools in order to achieve our objectives.

2.4 Bio-inspired Miniature Robotic Fish

As mentioned in the previous section, some groups are working on the development of actuated robotic fish for behavior studies with fish. Here, we will provide a short overview of some existing robotic fish platforms that could be used for this purpose.

Most of the existing robotic fish platforms are dedicated to bio-inspired robotic research (Du et al. 2015; Bandyopadhyay 2005). There are often new techniques proposed for the actuation of the fins; for instance, in Wang et al. (2008) and Chen et al. (2010), shape-memory alloys and electro-polymers are used respectively to actuate the fins.

A group has also worked on the design of an autonomous soft robotic fish (Marchese et al. 2013 (Fig. 2.3d). Their 34-cm long robot was self-propelled and able to reproduce the tail motion of fish effectively. However the robot is still too large for studies with small species of fish.

In Takada et al. (2010), a robotic fish 10 cm in length was designed with a DC motor to actuate the tail. The robot was also powered using fuel cells, targeting long duration autonomous exploration in flooded areas (Fig. 2.3c). This robot provides an interesting low-scale design, but, as far as we know, no further developments were made on this design since several years.

Some commercialized robotic fish have also appeared in recent years. For instance the *Robofish* toy (Zuru, China) is very popular and was the fastest selling toy in 2013 (Rob 2013). This robot is composed of a battery, an electro-magnet to actuate the caudal fin and some miniature electronics (Fig. 2.3a).

Another example of a commercialized robotic fish is the *Jessiko* (Robotswim, France) (Jes 2016) (Fig. 2.3b). *Robotswim* is a start-up created in 2009 with the main goal to bring artificial life into aquariums and pools. Jessiko is 22 cm long, can reach speed up to 20 cm s^{-1} and can form swarms based on knowledge of the other robots' position. It is composed of two actuators, the beating tail and a flap near the fish head. A sophisticated tracking system with receiver located on the corner of the aquarium allows a closed-loop control in positioning the robot swarm.

Unfortunately, the scales and dynamics of these products and these prototypes are an issue for this project, as they do not reach the average size and typical motion of the zebrafish. However, these designs suggest interesting approaches for the conception of robotic fish.

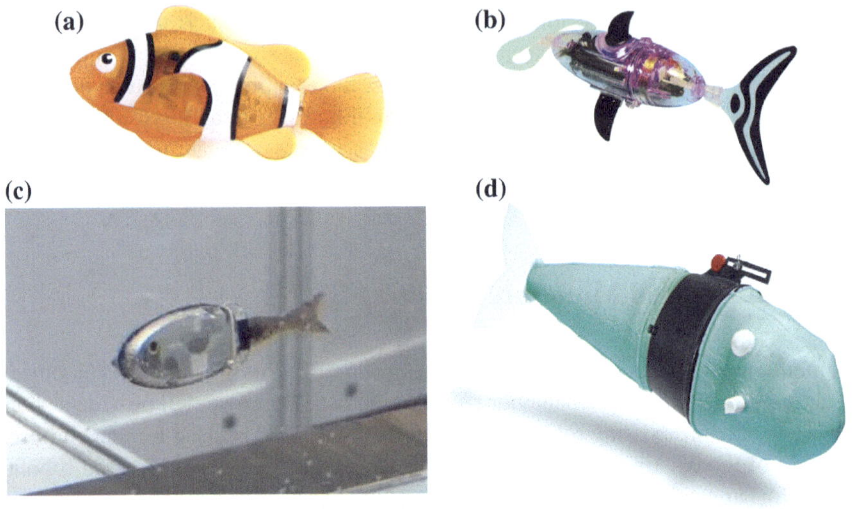

Fig. 2.3 **a** Robofish toy. This robot is equipped with small batteries, a motor composed of a coil and a magnet, and a small electric circuit. It is able to beat its tail with different frequencies. The robot can move autonomously, but its motions are not controlled in a closed loop. **b** Jessiko robot designed mainly for entertainment that can move autonomously in all directions. **c** A 10-cm long fish robot equipped with a DC motor to actuate the tail and powered by fuel cells (Takada et al. 2010). **d** A robotic fish that can move autonomously underwater by reproducing fish body motions (Marchese et al. 2013)

2.5 Our Contribution to the State of the Art

Concerning the fish-robot interaction studies, there are currently no groups that provide a solid multi-robot platform for the study of mixed societies of fish and robots. The authors in Landgraf et al. (2013) and Landgraf et al. (2016) claimed that their platform allows the simultaneous operation of multiple robots; however, none of their experiments were performed using multiple robots. The system of Landgraf et al. (2013) and Landgraf et al. (2016) also did not allow long duration experiments due to the fact that the battery onboard the robot needed to be changed. The robotic system provided by the group of Porfiri (Kopman and Porfiri 2013) did not allow a shoal of multiple robots to follow a fish shoal. While the robotized lures showed promising results, however, their size and their biometrics cues are not enough similar to zebrafish in our opinion.

In parallel to the development of the hardware, different approaches have been proposed to control the movement of the robotic fish used in behavioral studies (see Table 2.1). Most of these approaches involve a predetermined trajectories without taking into account the response of the fish. Such methods are relevant to investigate the preferences of fish for particular features but cannot be applied to integrate a shoal of fish with an artificial agent. On the other hand, closed-loop interactions that use the current positions of the fish to determine the next move of the robots have also been used for self-propelled robots but mostly for fish lures moved by small robots under the tank. This method requires the tracking in real-time of the agents (fish and robot) and a decision-making algorithm to control the trajectory of the robots. In the experiments reported in the literature, the robots using such a closed-loop control are programmed to follow the centroid of the group of fish or, in a more advanced version, to follow the centroid then aim towards a specified target (Landgraf et al. 2016). However, this algorithm does not ensure that the robots will join and follow the group of fish, as the robots need to anticipate the future position of the fish; otherwise, it will mostly be delayed compared to the collective movements of fish.

Concerning the development of a small robotic-fish device, there are currently no groups working with fish-like robots within the size range of zebrafish. Also, the current technology seems to not allow the development of a miniature and autonomous robotic fish with the same size range as zebrafish.

Thus, due to the state of the art at the beginning of this dissertation, in order to achieve the goal of this thesis, which is to realize experiments with mixed societies of zebrafish and robots, we decided to create our own robotic device and control infrastructure to achieve a closed-loop control of multiple robots, following models that reproduce the fish shoaling movements and locomotion patterns. This will contribute to fill a gap in the state of the art in fish-robot interaction studies.

Chapter 3
The Zebrafish *Danio rerio* as a Model Animal for Animal-Robot Interaction Studies

3.1 Summary

This chapter introduces the model organism that was used in the presented work, i.e, the wild-type AB zebrafish *Danio rerio* (Fig. 3.1). We will first present the main biological characteristics of the zebrafish and explain why it is widely used nowadays as a model animal in different scientific topics. Some descriptions of the zebrafish behavior observed during this study inside our experimental infrastructure and the facility created to house zebrafish in our laboratory will also be shown.

3.2 The Zebrafish Biology

There is a lot of literature on the biology of the zebrafish, but perhaps one of the most complete and famous references is the work of Rowena Spence, which studied many aspects of the zebrafish in its wild life conditions (Spence et al. 2008). First, the zebrafish is a freshwater fish that originated in the southeastern Himalayan region, mostly found in freshwater ponds and small canals in countries such as India or Bangladesh. It is a very robust fish that can adapt to many types of water conditions, with temperatures from 6° in winter to 38° in summer. Zebrafish are omnivorous and live in groups both in their natural environments and in laboratory conditions. Female zebrafish can lay about 100–200 eggs every two or three days. Individuals rarely exceed 40 mm in standard length (from the tip of the snout to the origin of the caudal fin).

F. Bonnet and F. Mondada, *Shoaling with Fish: Using Miniature Robotic Agents to Close the Interaction Loop with Groups of Zebrafish Danio rerio*, Springer Tracts in Advanced Robotics 131, https://doi.org/10.1007/978-3-030-16781-3_3

Fig. 3.1 One of the wild-type zebrafish *Danio rerio* that was raised and used for experiments in our laboratory during this research

3.3 The Zebrafish as a Model Vertebrate

One of the pioneer in the use of zebrafish as a model in scientific researches was George Streisinger (Streisinger et al. 1981). Legend has it that in his laboratory at the Institute of Molecular Biology in the University of Oregon (USA), Streisinger had a huge aquarium composed of several species of freshwater fish. One day, he left for the holidays and forgot to ask someone to feed and take care of the animals during his absence of several weeks. When he came back from holidays, he found all the fish dead in the aquarium, except for the zebrafish, which survived. He realized that this fish was thus robust enough to be used in research and decided to orient his genetics studies mostly on the zebrafish. He established the first protocol on zebrafish scientific studies, and several years later, his first publication on his success in cloning the zebrafish came out (Streisinger et al. 1981).

Since the work of Streisinger, the zebrafish has become a common model organism in genetics (Norton and Bally-Cuif 2010). In 2013, the zebrafish genome was fully sequenced to serve as a reference tool for biologists (Howe et al. 2013).

3.4 Stimuli Perceived by the Zebrafish

The zebrafish possess the same sensory organs as most of the Teleost fish: smell, taste, vision, audition and mechanosensation (Perry et al. 2010). The mechanosensing is done through the lateral line, an organ specific to the fish that is sensitive to mechanical stimuli, such as water movements, and can help the fish detect the motions of prey, predators, and also conspecifics (Dykgraaf 1933). It is also suggested that the shoaling behavior of teleost fish is mostly done using the sensory information from the vision organ and the lateral line (Pitcher and Parrish 1993). Other sensing organs are used for other activities such as reproduction, food foraging, etc. (Ioannou 2016). Thus, in our specific case, as we need to design a robot that can be perceived as a conspecific inside of a group, it should generate mostly similar stimuli as those perceived by the vision and lateral line organs of fish.

3.5 The Advantage of Using Zebrafish Compared to Other Animals

The zebrafish is not only a very resistant and easy to breed fish; it has also many other advantages that make it an interesting model organism for scientific studies, especially in developmental genetics and physiology.

- A vertebrate: The zebrafish is a vertebrate and thus shares about 70% of its genetic material with humans (Howe et al. 2013).
- External embryo development: Unlike the mouse or some other laboratory animals, the zebrafish embryo develops inside eggs fertilized outside the parents, and the eggs can be collected few seconds after fertilization, and the entire embryo development can be observed.
- Transparent embryo: On the first days of the embryo development, the cells are transparent. The pigmentation appears later in the development, and thus the organ development can be observed using markers, as shown in Fig. 3.2.
- Large number of embryos: One female can lay between 100 and 200 eggs per mating, and thus the number of samples is more than 10 times larger than for mice; moreover, the female is able to lay this amount of eggs every 2–3 days.

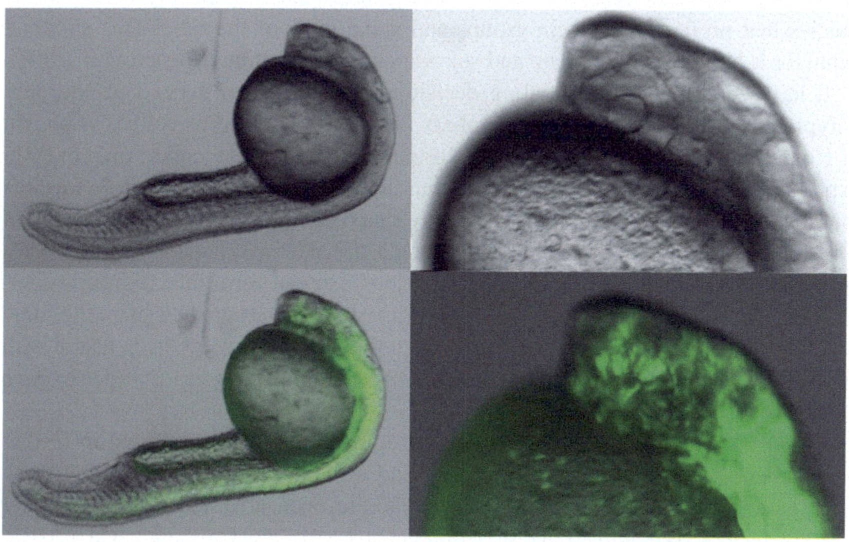

Fig. 3.2 Cells traceability using the zebrafish during embryonic development. Zebrafish eggs were collected right after fertilization. An mRNA marker was injected into one cell a few hours post-fertilization, when only four cells could be identified on the eggs. These pictures were taken one day after the injection. In both pictures, the marker can be identified, which helps us to trace how the injected cell developed inside the embryo. These experiments were performed at the University of Geneva in the department of genetic medicine during a class on the housing and care of laboratory animals

- Embryos are easy to manipulate: The egg is quite robust and easy to manipulate with a pipette or other tools.
- Cost/animal: Zebrafish can be raised inside rather small aquariums that can afford a large density of animals inside the laboratory. They are easy to breed because males and females can be raised together. They are also very robust and do not require a particular diet.
- Regenerative properties: Some parts of the zebrafish can regenerate, which is not the case in many other animal species, such as humans. For instance, zebrafish that are blinded can recover their vision after a few months (Gemberling et al. 2013), the heart of zebrafish will regenerate even after burns in some areas as shown in Poss et al. (2002) and Chablais et al. (2011), and the tail will regenerate after being cut (Poss et al. 2003).

3.6 Zebrafish Collective Behavior Under Laboratory Condition

Our study involves experiments with mixed societies of robots and zebrafish under laboratory conditions. In the past decade, the collective behavior of zebrafish under laboratory conditions has been the subject of several studies. The zebrafish is a social species that prefers to swim in groups and shoal most of the time. This shoaling behavior is believed to be innate and starts soon after hatching (Spence et al. 2008).

It has been shown that shoaling decisions in fishes often depend on the phenotypic characteristics of group members (such as body size or color), as shown in Hoare et al. (2004) and Wong and Rosenthal (2005). For the case of zebrafish, experimental results have demonstrated that individuals show preferences towards their own conspecifics, with the color, body shape and stripes playing an important role, as demonstrated in Rosenthal and Ryan (2005) and Saverino and Gerlai (2008). However, it is also known that some gregarious fish can form mixed groups composed of different fish species (Pitcher and Parrish 1993), when they share the same food diet (Kleinhappel et al. 2016). Although this effect has never been tested with zebrafish, it may not be crucial to create an exact zebrafish replica in order to have an infiltrated agent inside the fish shoal.

Under our laboratory conditions, in the setups that will be described in further chapters composed of a white background and walls to constrain the motion of the fish, the zebrafish usually tend to follow walls in open area and aggregate under floating devices, as demonstrated in Séguret et al. (2016) and Collignon et al. (2016). In the case of a constrained area, such as corridors 10 cm in width, they tend to swim fast along the corridor to reach its end, and, in the case of an infinite corridor, they sometimes make U-turns. Regarding the shoaling behavior, we observed that the zebrafish always have the tendency to stay in group; however, depending on the number of individuals and the size of the arena, the homogeneity of the shoal may vary, with, for instance, fish in experiments with more than 10 individuals sometimes forming two or more shoals.

Recent studies have also shown that the strain of the zebrafish can also affect the collective behavior of the zebrafish shoal in environments composed of an open tank with landmarks where the zebrafish generally aggregate (Séguret et al. 2016). In the work presented in this study, we will show only results of experiments with the same strain of zebrafish (AB). However, the collective behavior observed in this study might not be reproducible depending on the strain of zebrafish used.

3.7 Zebrafish at the EPFL

For the experiments performed, we used in total 110 wild-type AB zebrafish *Danio rerio*, with short fins. These zebrafish were acquired in a pet shop and were stored in two 60-L housing aquariums (Fig. 3.3). The average total length of our zebrafish was 40 mm. We kept the fish under laboratory conditions: 27 °C, 500 μS salinity with a 10:14 day:night cycle. The fish were fed twice a day using a food distributor with commercial food. The water pH level was maintained at 7.5, and Nitrites (NO^{-2}) were below 0.3 mg/L. The zebrafish were raised in a room separated from the one in which the experiments were conducted.

It was demonstrated that the zebrafish have a preference for enrichment, with an increase of their social behavior (Schroeder et al. 2014). Thus, we added enrichment in the housing aquarium that consisted of plastic plants, cladophora balls, gravel, rocks and aquatic snails.

Fig. 3.3 Fish facility in our research laboratory. Two 60-L tanks were used to host the zebrafish. The enrichment consisted of plastic plants, cladophora balls, gravel, rocks and aquatic snails

3.8 License

Most of the experiments involving zebrafish that are described in this work were performed in our laboratory at the Ecole Polytechnique Fédérale de Lausanne in Switzerland. They were conducted under the authorization N°2778 delivered by the Department of Consumer and Veterinary of the Canton de Vaud (Switzerland) after submission to the state ethical board for animal experiments.

Some experiments were performed by our partners in the ASSISIbf project at the University Paris Diderot (France). They were conducted under the Buffon Ethical Committee (registered to the French National Ethical Committee for Animal Experiments #40) after submission to the French state ethical board for animal experiments.

Chapter 4
FishBot, the Fast Miniature Wheeled Mobile Robot

4.1 Summary

In this chapter, we present one of the two robotic devices that was developed during this thesis: the FishBot, a miniature wheeled mobile robot, which main purpose is to transmit motion to a fish lure inside an aquarium (Fig. 4.1).

We will first describe the requirements that were set to perform biological studies with zebrafish involving this robot, then show the different solutions and versions of the robot that were developed throughout this thesis and, finally, the validation of the robot's final version to perform the experiments with zebrafish.

This chapter is based on the publication *Development of a Mobile Robot to Study the Collective Behavior of Zebrafish* (Bonnet et al. 2012) and partly on the publication *A Miniature Mobile Robot Developed to be Socially Integrated with Species of Small Fish* (Bonnet et al. 2014) with slight modifications to fit the present dissertation.

4.2 Requirements and General Descriptions

The design of a robot dedicated to experiments with animals have to originate from relevant sensory modalities and behaviors of the animal under study (Mondada et al. 2012). Hence, a good understanding of animal biology and behaviors, both on the individual and collective levels, must precede the robot design process.

As it was previously mentioned, one of the main requirement for this thesis was to design a robot capable of interacting with zebrafish. There were two types of requirements for such experiments, the ones that were set by the animal itself, as our robot had to mimic several physical aspects of the animal under studies in order to be perceived as a conspecific, and also the ones that were set by the experimenter

F. Bonnet and F. Mondada, *Shoaling with Fish: Using Miniature Robotic Agents to Close the Interaction Loop with Groups of Zebrafish Danio rerio*, Springer Tracts in Advanced Robotics 131, https://doi.org/10.1007/978-3-030-16781-3_4

Fig. 4.1 The FishBot, a miniature wheeled mobile robot that is able to steer lures so that they achieve fish movements and shoaling underwater, while being continuously powered

that would use the device to perform the experiments involving living animals; these experiments were driven by certain protocols that were set by biologists in order to obtain very robust measurements.

4.2.1 Biological Specifications

A list of specifications was created at the beginning of the project for the robotic fish lure. The specifications were based on the existing scientific literature on zebrafish and observations made on our own animals. For instance, thanks to the tracking system that will be further described (see Chap. 7), we could extract the individual speed and acceleration of our own zebrafish to adapt the specifications of our design during the development phases of the robots.

Based on all these metrics, we drew up the following specifications for the first version of the robot:

- **Dynamic**: The robot should have the following dynamic capabilities: a maximal speed of at least $0.5\,\mathrm{ms}^{-1}$ and maximal acceleration of $2\,\mathrm{ms}^{-2}$. The speed and acceleration parameters were selected based on the available experimental results obtained with our own animals.
- **Size**: The dimensions of the robotic fish should correspond to the dimensions of the real fish: total length 45 mm, maximum body height 10 mm, maximum body width 5 mm. The robot had to be designed taking into account that it would be used in experiments involving multiple robots, where robots can approach each other very closely (1–2 cm). The robot should also have a visual appearance as close as the one of a zebrafish.
- **Communication**: The robots have to provide wireless communications so that their behaviors can be modified and automatically adapted during an experiment. Indeed, the robot should be control in a closed-loop while adapting to the fish behavior using computationally intensive models, thus the control would not be fully embedded on the robot.

- **Long-duration experiments**: The behavior of zebrafish varies overtime, as it was observed in our own experiments. For instance, when placed into a new environment, the zebrafish could show stress behavior during several minutes. This behavior will then disappear and followed by an exploration of the environment. In Chap. 2, we showed that most of the researchers in this field are measuring the effect of lures on the zebrafish behavior using experiments relatively short in time, which is partially due to the low autonomy of the robotic system. Thus, a system able to work continuously over long periods of time could be beneficial for the biologists. It was set as a requirement for our robotic system to be able to work continuously without any human interventions.

4.2.2 Solutions Summary

Of course, the immediate reaction of an enthusiastic engineer in robotics would be to design a small autonomous underwater biomimetic robot able to reproduce fish locomotion to follow them thanks to a closed-loop control. Unfortunately, even though we did not lack of enthusiasm at the beginning of the project, we realized that, during the time this work was done and due to the specifications, this was not achievable. Indeed, in terms of speed and acceleration required, it would not be possible to design with the current technology an autonomous robotic fish with the required size that could be controlled in a close-loop to achieve such movements autonomously in the water. Therefore, we decided to start with the design of a robot moving outside an aquarium.

The external robot needed to provide very dynamic movements in order to mimic the ones of the zebrafish. The use of wheels instead of other types of locomotion modes was the most evident solution for this, as it is possible to create a shoal of robots moving autonomously in different directions. Indeed, the solutions that are suggested in Faria et al. (2010) or Kopman and Porfiri (2013) using two-dimensional moving platforms or robotic arms were not convenient for multi-robot experiments.

As the zebrafish almost never produce lateral displacements as it will be demonstrated in Chap. 8, a non-holonomic locomotion for the wheeled robot was acceptable. Thus, we selected a two-wheel differential-drive configuration with two additional ground contact points used for stability. Such a configuration offers a high maneuverability to the robot which can also rotate while staying at the same position, as the zebrafish do. The robot would be magnetically coupled with a module moving in the water, and the movements produced by the robots would be transferred to this module so that it mimics the zebrafish movements underwater (Fig. 4.2). This solution allowed the robotic system to achieve the required speed and acceleration, thanks to the robot moving outside the aquarium, while having a lure inside that could have the same size and visual appearance as the zebrafish.

This chapter will describe the wheeled mobile robot that was design to move outside the aquarium: the FishBot. The design of a robotic lure that can be

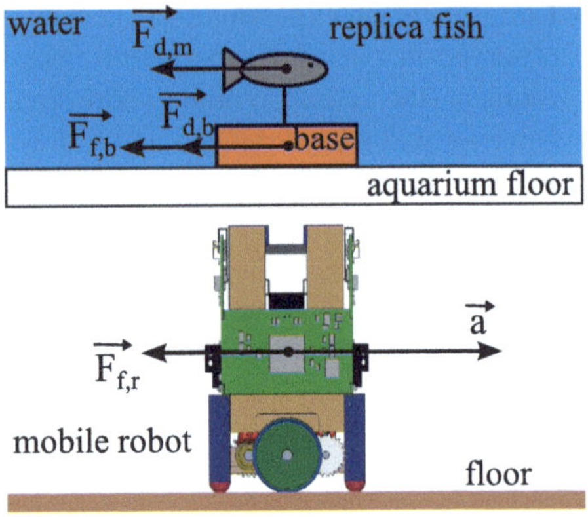

Fig. 4.2 Diagram of forces acting upon the mobile robot FishBot and the fish replica module when the robot accelerates. Drag forces $F_{d,m}$ and $F_{d,b}$ act upon the fish replica and upon its base, and friction forces $F_{f,b}$ and $F_{f,r}$ are exerted by the tank floor on the base of the fish replica module and by the floor on the ground contact points of the mobile robot

magnetically coupled with the FishBot and that can also reproduce the body movements of zebrafish inside the aquarium will be described in Chap. 5.

4.3 Actuation of the First FishBot Version

To select suitable motors for the robot, we estimated the required rotation speed ω_m and torque T_m as

$$T_m = \frac{F_r R_w}{2r}, \tag{4.1}$$

$$\omega_m = \frac{r V_r}{R_w}, \tag{4.2}$$

where F_r is the force needed to accelerate the robot at the maximal acceleration ($2\,\text{ms}^{-2}$), R_w is the radius of the wheels, r is the reduction between the motors and the wheels and V_r is the maximal speed ($0.5\,\text{ms}^{-1}$).

The forces acting upon the system when the robot accelerates are represented on Fig. 4.2. Hence, the force F_r needed to accelerate the robot is

$$F_r = Ma + F_{f,b} + F_{f,r} + F_{d,m} + F_{d,b}, \tag{4.3}$$

where M is the mass of the robot and the fish replica module together that was estimated at 180 grams. The friction forces ($F_{f,b}$ and $F_{f,r}$) have been estimated using the static friction formula $F_f = \mu_s P$, where P is the weight of the body and μ_s the static friction coefficient. The drag forces ($F_{d,m}$ and $F_{d,b}$) on the fake fish and the mobile robot have been computed as

$$F_d = C_d A \left(\frac{\rho V_0^2}{2}\right). \tag{4.4}$$

where A, ρ and V_0 are the drag area, fluid density and free-stream velocity measured relative to the object, respectively. The drag coefficient C_d has been estimated using engineering tables from Crowe et al. (2010).

Two DC motors (Maxon, Switzerland) were preselected due to their small diameter (8 mm) and high nominal torque (0.616 mNm). The magneto resistive encoders of these motors (Maxon encoder MR Type S 100CPT) provide 6400 position readings per wheel's round thanks to the 16:1 reduction ratio, which guarantees a good precision for speed and position control. Maxon 16:1 reduction ratio gearboxes (Maxon Planetary Gearhead GP 8A 16:1) were combined with the motors to multiply their torque and a wheels' diameter of 20 mm was chosen in order to obtain the best trade-off between the maximal torque T_m and the maximal rotational speed ω_m of the wheels in order to achieve the required movements.

As it is shown on Fig. 4.3, bevel and spur gears were used to transmit the motion to the wheels. Once all the components were selected, we did an estimation of the final torque taking into account yield and inertia of the transmission in order to validate the choice of the motors, gears, wheels diameters and transmission.

We used the following relation between the torque T, inertia of the wheel I_{fW} and angular acceleration α_w of the wheels:

$$T = I_{fW}\alpha_w \tag{4.5}$$

Here, the inertia seen from the wheel I_{fW} is computed using all inertia and yield of the transmission

$$I_{fW} = I_w + \eta_R\eta_b\eta_s r^2 I_m + \eta_R\eta_b\eta_s r^2 I_R + \eta_b\eta_s I_b + \eta_s I_s, \tag{4.6}$$

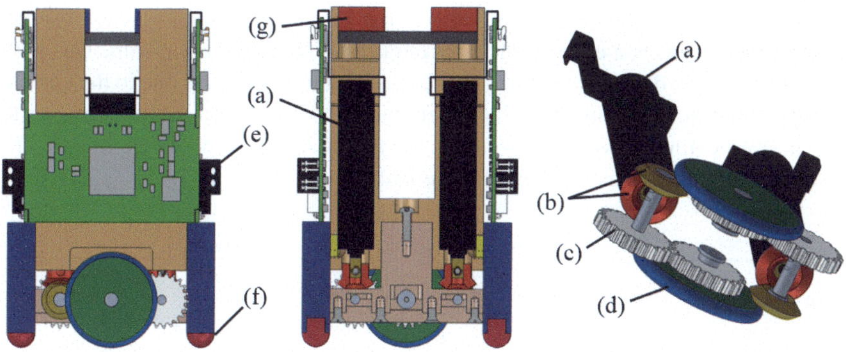

Fig. 4.3 Design of the first version of the FishBot. Left: the FishBot seen from the side with its Printed Circuit Board (PCB) (in green) mounted. Middle: sectional view of the FishBot. Right: View of the transmission of the FishBot with bevel gears that were used for the transmission between the vertical motors and horizontal axes, and spur gears were used to transmit the motion to the two independent wheels. **a** Assembly of the motor, the gearbox and the encoder. **b** Bevel gears. **c** Spur gears. **d** Wheels. **e** Infrared proximity sensors. **f** Skates. **g** Magnets

where I_w, I_m, I_R, I_b, and I_s are the inertia of the wheel, motor, gearbox, bevel gear and spur gear respectively. η_R, η_b, and η_s are the yield of the gear, bevel gear and spur gear respectively, and r is the gearbox reduction. We assumed here that the robot's wheels do not slip on the ground.

We could then obtain the angular acceleration α_w of the wheels:

$$\alpha_w = \frac{r\eta_R\eta_b\eta_s T_m - T_w}{I_{fW}}, \tag{4.7}$$

where T_m and T_w are the torque of the motors and wheels, respectively.

This estimation showed that the mobile robot could theoretically move the fish replica module with a maximum speed of 0.87 ms^{-1} and acceleration 5 ms^{-2} at the motor nominal voltage of 6 V, thus validating the choice of the motors and the transmission.

4.4 Design and Tests of the First FishBot Version

We designed a first version of the FishBot (Fig. 4.3) integrating the motorization described in the previous section. Using this version, we ran a series of tests in the experimental tank that will be described in Chap. 6. Results of a sample test are presented on Fig. 4.4. Noise on data has been decreased using a Savitzky–Golay filter. In this test, the robots was programmed to reach the speed of 0.6 ms^{-1} with the shorter amount of time as possible.

As we can see from Fig. 4.4, the lure steered by the robot was able to achieve a speed of 0.57 ms^{-1} with the peak acceleration equal to 1.8 ms^{-2}. The demonstrated values were close to the ones of our own living zebrafish, and, as we can observe, the achieved speed even exceeds the ones required. The acceleration value was 10% lower than the one defined originally in the specification. This was due to the current settings of the motor controllers that limited the maximum current in the motor to prevent overheating. However, in general, the motorization based on the Maxon DC motors and encoders was validated for the FishBot.

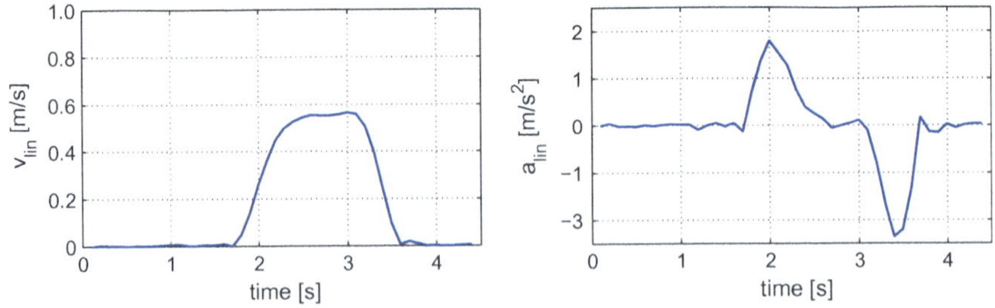

Fig. 4.4 Speed and acceleration profiles of the first version of the FishBot

4.5 Design History

Although some concepts of the first version could be validated, its major problem was the stability. Most of the structure was made of plastics, thus the center of mass of the robot was located too high. Additionally, the width of the wheels was too thin, thus most of the time, when reaching high speeds, the robot would fall during turns. Another issue was the robot's powering, which was made through cables. Thus, we designed new versions to integrate improvements in order to correct these issues.

Figure 4.5 shows the four designed versions of the mobile robot FishBot. A second version was designed involving a robot that was moving directly on the aquarium (upside down compare to the first version), using the magnetic attraction with the lure module to compensate the gravity force. The wheels were also shifted with an offset from the center, in order reproduce the dynamics more similar to the one of the zebrafish, and also to avoid the use of one level of spur gears inside the transmission. Poor results regarding the motion control of this version were obtained, mainly because the magnetic attraction forces were either too high which involved to much friction during the motion or too low which led to a poor stability of the robot. Therefore, the solution of having the mobile robot directly moving on the aquarium was abandoned for further versions. However, the powering of the robot using two electric conductive plates and brushes was tested on the second version, and it was demonstrated that the robot could be used for long experiments without any human intervention. This system was kept for the next versions. This powering system will be explained in detail in Sect. 4.6.2.

For the third version, the robot configuration was similar to the first version with the two wheels centered and the robot moving on a plate located under the aquarium. The structure with a chassis made of brass, which heavy mass lowers the robot's center of mass, guaranteed a better stability. The third version had the same transmission system as the first version. This third version offered very acceptable results in terms of dynamics and reliability, with experiments performed without any major failures

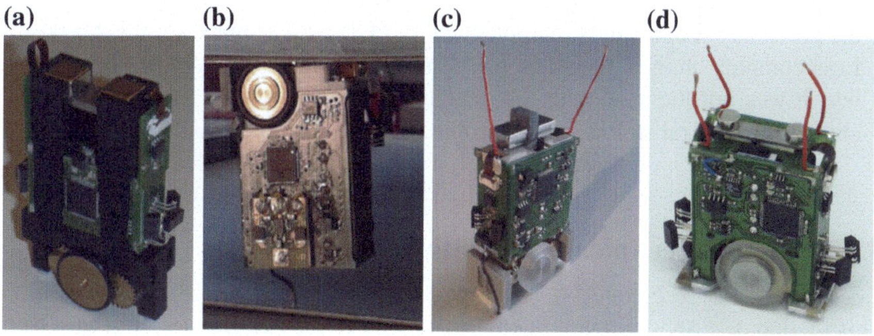

Fig. 4.5 Design history of the FishBot mobile robot's versions. **a** Version V1.0 prototyped in 2011. **b** and **c** respectively V2.0 and v3.0 built in 2013. **d** the version V4.0, which is the current version of the robot

of the robots. Two robots of this version were built, one for tests performed at the EPFL and one for the experiments in Paris. The latter was used as a demonstrator during the first Review Meeting of the ASSISIbf project at the end of March 2014, which could demonstrate that this robot can be used for experiments of interaction with zebrafish. These robots were tested during hundreds of hours. They were used to perform for instance the experiments in Bonnet et al. (2014) involving a robot moving among a small shoal of zebrafish that will be described in detail in Chap. 9.

Despite the very promising results obtained with the third version, several improvements, such as the size of the robot, production costs, reliability and user-friendliness had to be done on the mobile robot to obtain a version that could be easily produced in the perspective of a large production and easier maintenance that was required for the ASSISIbf project. Therefore came a fourth version of the robot, which is still the current version and the last one that will be developed in this project.

The new elements of the current version V.4.0 compared to the V.3.0 are:

- Worm gears for the motion transmission between the DC motors and the wheels: this solution offers a decrease of the robot's height as worm gears are used to transmit the motion from the vertical DC motors to the horizontal wheels axis and for the 16:1 reduction, thus the gearboxes, spur and bevel gears were no more needed. It also reduced the cost of the robot, as worm gears could be easily manufactured in our workshop and were less expensive than the Maxon gearboxes. Finally, the play inside the transmission was also reduced, due to reduction of gear levels, which offers a more accurate control of the wheel's motion for speed and position control.
- The power supply was set to 12 V instead of the 6 V that were used for the previous versions, which increased the quality of the brushes' contact with the conductive plates. It involved adding a DC/DC converter onboard each mobile robot as the input voltage of the power management chip was 5.5 V, and more powerful power sources for the case of multiple robot experiments.
- A supplementary Printed Circuit Board (PCB): Due to the size reduction of the mobile robot, the Bluetooth dongle used for the wireless communication between the robot and a computer could not be placed on the central board as it was done with the previous versions, thus a supplementary PCB was designed for this purpose.
- Less complex chassis: the Maxon DC motors could be screwed on the chassis, which was not the case with the gearboxes mounted on the motors. Thus, the chassis, the central structure of the robot, became easier to manufacture.

4.6 Detailed Technical Description of FishBot V.4

This section describes in detail the fourth version of the FishBot design (Fig. 4.6).

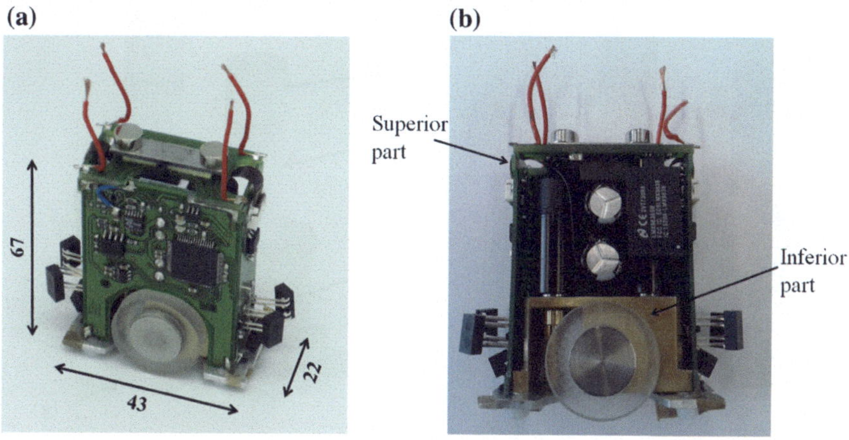

Fig. 4.6 Left: Dimensions of the fourth and current version of the FishBot mobile robot: The length is 43 mm, the width 22 mm and the height 67 mm which makes it the thinnest wheeled mobile robot used for this type of study. The mass of the mobile robot is 90 grams. Right: Side view of the mobile robot with the two parts of the structure defined

4.6.1 Mechanical Design

The current version of the FishBot is composed of two main parts (Fig. 4.6b): an inferior part that is composed of a chassis holding the two Maxon DC motors and the two wheel's axis, and the superior part that is composed of Printed Circuit Boards (PCBs) and on which the magnets are fixed to guide the lure module.

The inferior part of the current FishBot mobile robot is shown in details in Fig. 4.7. The chassis structure is made of brass, an easily machined metal with a high density in order to lower the mobile robot's center of mass to obtain more stable movements at high speed. Two Maxon DC motors are screwed on the chassis. The motors are equipped with magneto-resistive encoders. Worm gears made of brass are used for the transmission of motion and reduction. The spur gears are made of Polyoxymethylene (POM) to decrease the friction between the worm gear and spur gear, and thus increase the efficiency of the transmission. We manufactured the worm and spur gears ourselves, in order to obtain the most efficient characteristics for our design. We designed the worm gear with a double thread also to have a better efficiency, and kept the 16:1 reduction ration for the transmission as it was done for the previous versions.

For the worm gear, we chose a module m_n of 0.5, thus the pitch diameter of the spur gear was given by

$$Z_{spur} \cdot m_n = D_{p,spur} = 16\,mm \tag{4.8}$$

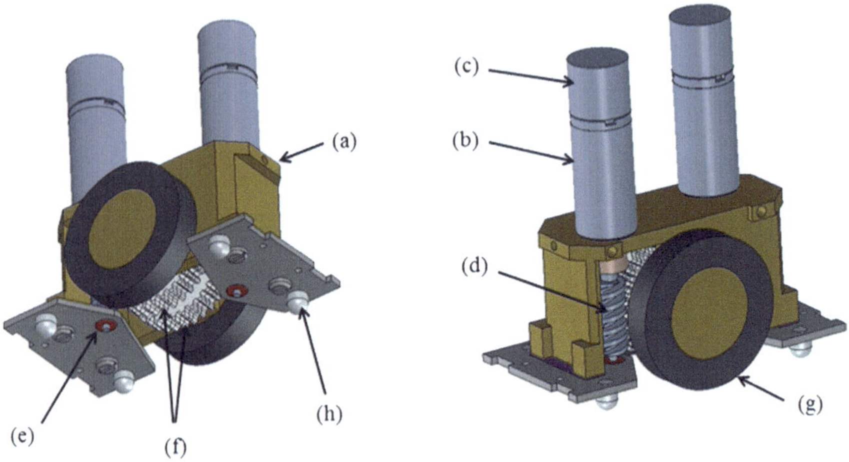

Fig. 4.7 The inferior part of the current version of the FishBot mobile robot. **a** The chassis structure. **b** Maxon DC motor. **c** Magneto-resistive encoder. **d** Worm gears. **e** Bearing. **f** Spur gears. **g** Wheel. **h** Skate

where Z_{spur} is the number of teeth of the spur gear, which is 32, given by the fact that the worm gear has a double thread and a reduction ration of 16:1. The total diameter of the spur gear is thus

$$D_{p,spur} + 2 \cdot m_n = 17\,mm \tag{4.9}$$

For the worm gear, we computed first the pitch p_w which is given for a two thread worm gear by

$$p_w = m_n \cdot \pi \cdot 2 = \pi\,mm \tag{4.10}$$

and the pitch diameter D_p

$$D_p = D_{ext} - 2 \cdot m_n = 3\,mm \tag{4.11}$$

with D_{ext} the external diameter that was set at 4 mm due to the mechanical constraints.

The lead angle is given by

$$\text{atan}\left(\frac{p_w}{\pi \cdot D_p}\right) = \text{atan}\left(\frac{\pi}{\pi \cdot D_p}\right) = \text{atan}\left(\frac{1}{3}\right) = 18.43^\circ \tag{4.12}$$

Bearings were fitted on two PCB plates that were screwed below the chassis in order to avoid the bending of the motors' shafts. The wheels' tires are made of polyurethane which soft texture offers a high adhesion with the ground and a good stability of the robot. Four Polytetrafluoroethylene (PTFE) skates are fixed below the chassis in order to serve with the two wheels as a third contact point on the ground.

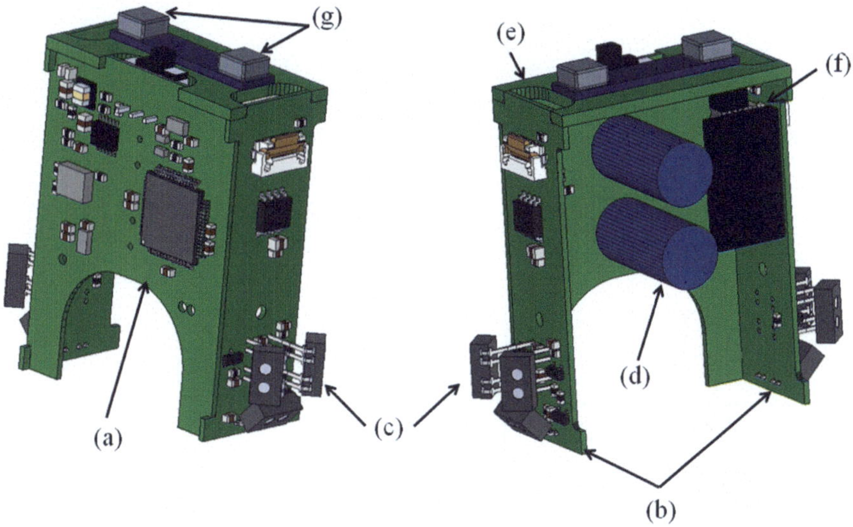

Fig. 4.8 The superior part of the current FishBot mobile robot. **a** The central board. **b** The two side boards. **c** Infrared proximity sensors. **d** Two SuperCaps. **e** The superior board. **f** Bluetooth module LMX 9838. **g** Magnets

The superior part of the current FishBot mobile robot is shown in details in Fig. 4.8. It is composed of four PCBs which are soldered together at an angle of 90 degrees to guarantee a stable connection between them. The central board, which is positioned on one side of the chassis using pins, contains the microcontroller and the power management system. The two side boards are identical in order to have a simplified design and assembly. The motor drivers, Infrared (IR) sensors and their drivers are placed on the side boards. The IR proximity sensors TCRT1000 are installed on the front and back of the robot in order to avoid other robots or the borders of the arena. The superior board connected with the three other boards is equipped with a DC/DC voltage converter and the Bluetooth module LMX 9838 for wireless connection. Magnets are placed on the top of the robot for magnetic coupling with the fish replica module.

4.6.2 *Electronic Design*

The electronic architecture of the mobile robot is presented in Fig. 4.9. The power supply is done through electric cables (brushes) that slip against two conductive plates situated under the aquarium's floor (positive) and on the support on which the robot is moving (ground). Two SuperCaps of 1 farad each with a dual ideal diode system are used to store power in case the brushes are not in contact with the conductive plates for at least five seconds at full speed in order to have a continuous powering of

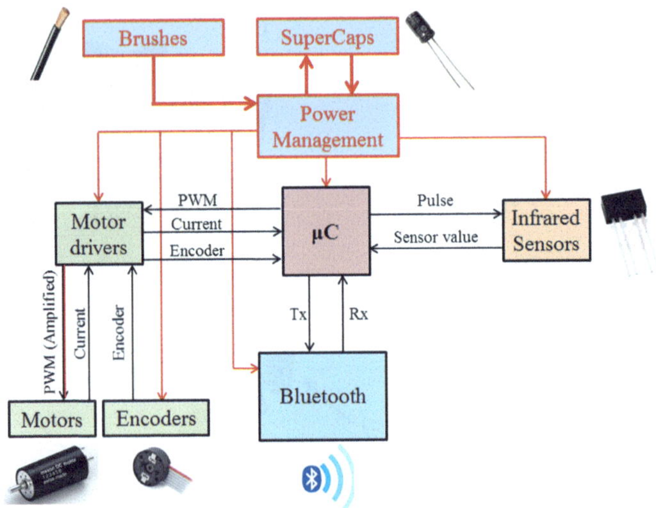

Fig. 4.9 Electronic architecture of the mobile robot. The microcontroller, a dsPIC33FJ128GP804, is monitoring the sensors and motors, and transmit data through a Bluetooth module. The brushes are used to power the system and load the SuperCaps, and when the contact between the brushes and the conductive plates is lost, the SuperCap can furnish the power for the whole system

the robot. The LTC4425 (Linear Technology, USA) is used to manage the powering of the mobile robot.

The microcontroller is a dsPIC33FJ128GP804 (Microchip, USA). This microcontroller can be reprogrammed through the same serial connector that is used for the Bluetooth connection. The Bluetooth device LMX 9838 is used to communicate between the mobile robot and the main computer, and is used for telemetry and control of the robot.

The IR proximity sensors TCRT1000 can detect obstacles up to a distance of 10 cm which is enough for obstacle avoidance even at high speed. The infrared emitted signal is pulsed and the signal measured is comparing the received pulsed signals in order to remove the noise and avoid the influence of the IR sensors from the other robots.

The motors are driven using SI9986CY H-bridges and the encoders value in the form of a pulsed signal as well as the motors current are retrieved by the microcontroller for current, speed and position control of the motors.

Six monochrome LEDs are fixed underneath the robot, three on the front and three on the back (see Fig. 7.9). This allows the robots to be better tracked from below the support on which they are running. This method will be described in more detail in Chaps. 6 and 7. One LED color is attributed to each FishBot for identification.

4.6.3 Firmware

The firmware is implemented in C language on the dsPIC microcontroller. It is based on two existing libraries: Molole and Aseba.

Molole (Mobots low level library) is a collection of low-level functions and drivers for the dsPIC processor family aimed at robotics (Mol 2016). This library is used to drive the different peripherals.

On top of that, an Aseba Virtual Machine (AVM) (Magnenat et al. 2011) is running onboard each mobile robot. Aseba is an event-based architecture for real-time distributed control of robots (Ase 2016). Running Aseba scripts inside each virtual machine enable us to reprogram the low-level behavior of each robot without flashing each time their microcontrollers (Fig. 4.10). It allows the users of the FishBots to reprogram easily their low-level behaviors remotely.

All the robots embedding an AVM can be connected together on the same Aseba network and communicate through events. Aseba integrated with D-Bus allows access to each robot from high-level applications that can be codded in languages, such as C++ or Python, using a software hub called Medulla (Magnenat and Mondada 2009).

Aseba was also chosen for its event-based architecture, which leads to very reactive behaviors of the robots. Indeed, the zebrafish is a very reactive animal, which moves fast with abrupt turns. Thus, an architecture based on nodes that can exchange information through events and react rapidly was required for the system.

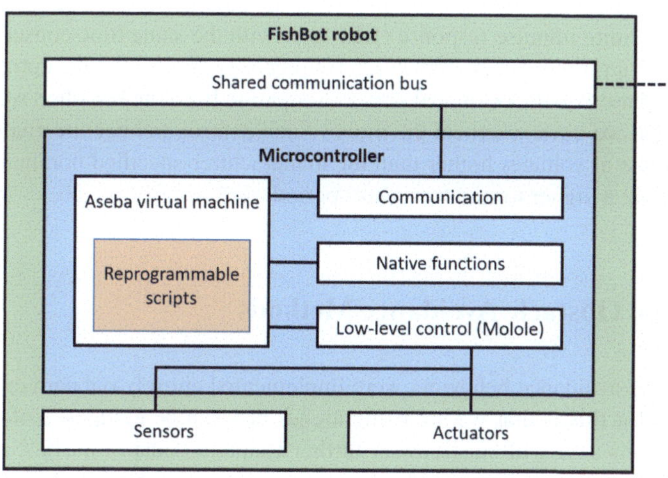

Fig. 4.10 Firmware architecture in the FishBot microcontrollers (Magnenat et al. 2011). The microcontroller code is split into two parts: First, sensor readings, actuator low-level control, and the communication layer are implemented in native code on the microcontrollers. This allows real-time, interrupt-driven handling of hardware resources. Second, application-specific programs that control the event emission and reception policy run in a AVM on the microcontrollers. They are compiled out of a simple scripting language, which provides the necessary flexibility to allow the application developer to implement the event-based behavior (Magnenat et al. 2011)

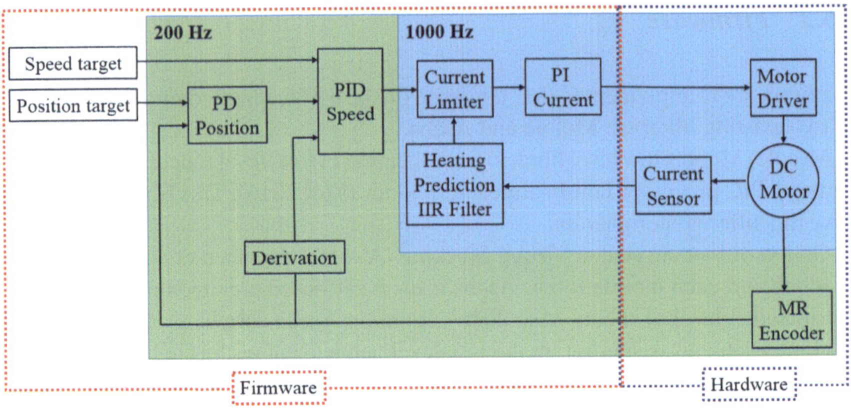

Fig. 4.11 Control architecture of the FishBot motors. A PI current controller is running at 1 KHz. The motor encoders retrieve the position of the motor axis that is used in a PID speed control loop and PD position control loop at 200 Hz. The user can choose between only current, position or speed control of the motor

In order to increase the capabilities of the robot to mimic the locomotion of zebrafish in terms of speed and acceleration, a cascade controller is implemented for the low-level control of each motor (Fig. 4.11). Each motor's torque is controlled by a Proportional–Integral (PI) controller running at 1 KHz, which is itself controlled in speed by a Proportional–Integral–Derivative (PID) controller or in position by a Proportional–Derivative (PD) controller using the motor encoders measurements as inputs. An infinite impulse response (IIR) filter with the same time constant as that of the motor runs on the microcontroller. By precisely controlling the current inside each motor, based on the estimated power dissipation from the IIR filter, we are able to ensure that we never overheat the motor. Such control architecture enables us to use the motors at voltages higher than the manufacturer-specified nominal voltage, thus providing a higher torque for a short period of time.

4.7 Local Obstacle Avoidance Methods

Two collision avoidance behaviors were implemented entirely onboard each robot. The reason for this is that we are using arenas that can be composed of corridors or small rooms as it will be shown in further chapters (Chap. 6). Thus, a reactive and reliable obstacle avoidance behavior was required at the low level of the control architecture to avoid the collision of the robots with boarders of the arena or other robots, that, otherwise, could generate a failure of the experiment with a FishBot blocked or a decoupling between a fish replica inside the aquarium and a FishBot.

We implemented these local obstacle avoidance mechanisms based on the measure obtained from the IR proximity sensors of the FishBot. Once an obstacle is detected, a corresponding event is sent over the Aseba network to inform the high-level software, so that the user can be informed about the event, and the high-level controllers will react depending on the gravity of the event. Two types of obstacle avoidance methods were implemented, one to generate a robust escaping behavior, and one based on a Braitenberg vehicle technique to allow the robot to avoid smoothly the walls of the arena (Braitenberg 1986).

4.7.1 Turn and Avoid

This solution stops and rotates the FishBot when one or two IR sensors detect the presence of an obstacle. The parameters of this algorithm, such as the speed of rotation, the duration of the rotation and the threshold to determine that an obstacle was detected by the IR sensors can be adapted depending on the experimental conditions (Figs. 4.12, 4.13 and 4.14).

The advantage of this algorithm is that the robot will always escape from a region with obstacles. The disadvantage is that the robot produced movements that are not biomimetics. For instance, the zebrafish have a tendency to follow the walls of the arena that we use, and, with this method, the robot will not mimic this behavior close to walls.

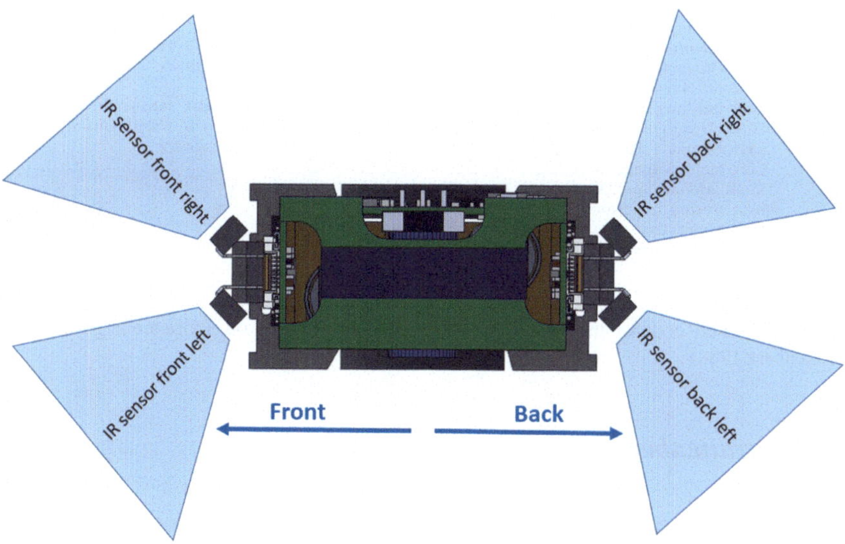

Fig. 4.12 Top view of the FishBot with the four IR proximity sensors that are used for the local obstacle avoidance

```
ThreshSensor=Threshold value to detect obstacle
SpeedAvoid
Store Previous Motor Speed
if IR front left sensor > ThreshSensor and IR front right sensor < ThreshSensor then
    emit Obstacle[IDControl]            #send an event on the Aseba network
    Left Motor=–SpeedAvoid
    Right Motor=+SpeedAvoid             #This make the robot turn
    Wait during a period of Time
    Left Motor=Last left motor speed
    Right Motor=Last right motor speed
elseif IR front right sensor > ThreshSensor and IR front left sensor < ThreshSensor then
    emit Obstacle[IDControl]            #send an event on the Aseba network
    Left Motor=+SpeedAvoid
    Right Motor=–SpeedAvoid             #This make the robot turn
    Wait during a period of Time
    Left Motor=Last left motor speed
    Right Motor=Last right motor speed
elseif IR front left sensor > ThreshSensor and IR front right sensor > ThreshSensor then
    emit Obstacle[IDControl]            #send an event on the Aseba network
    Left Motor=–SpeedAvoid
    Right Motor=–SpeedAvoid             #This make the robot go back
    Wait during a period of Time
    Left Motor=Last left motor speed
    Right Motor=Last right motor speed
end
```

Fig. 4.13 Pseudo code of the obstacle avoidance strategy Turn and Avoid

```
ThreshSensor=Threshold value to detect obstacle
CoeffAvoid=coeffient that will divide the IR sensor value
Store Previous Motor Speed
if IR front left sensor > ThreshSensor and IR front right sensor < ThreshSensor then
    emit Obstacle[IDControl]            #send an event on the Aseba network
    Left motor speed=Left motor previous speed – (IR front left sensor/CoeffAvoid)
    Right motor speed=Right motor previous speed
elseif IR front right sensor > ThreshSensor and IR front left sensor < ThreshSensor then
    emit Obstacle[IDControl]            #send an event on the Aseba network
    mot1.pid.target_speed=Left motor previous speed
    mot2.pid.target_speed=Right motor previous speed + (IR front right sensor/CoeffAvoid)
elseif IR front left sensor > ThreshSensor and IR front right sensor > ThreshSensor then
    emit Obstacle[IDControl]            #send an event on the Aseba network
    Left Motor=–SpeedAvoid
    Right Motor=–SpeedAvoid             #This make the robot go back
    Wait during a period of Time
    Left Motor=Last left motor speed
    Right Motor=Last right motor speed
end
```

Fig. 4.14 Pseudo code of the obstacle avoidance strategy based on a Braitenberg algorithm (Braitenberg 1986)

4.7.2 Braitenberg

The Braitenberg vehicle principle is controlling the motion of the motor by virtually linking the sensor with the motorization (Braitenberg 1986). The speed of the motors is thus varying depending on the sensor values obtained. We implemented this algorithm on the FishBot, in order to avoid the obstacle more smoothly, which allows the robot to follow the walls of the arena for a certain amount of time depending on the commands that are sent from the high-level control application.

Compared to the Turn and Avoid algorithm, the Braitenberg algorithm offers an obstacle avoidance behavior more close to the behavior of zebrafish. However, the robots might get stuck depending on the environment. Thus, it is less robust to guarantee a safe obstacle avoidance than the Turn and Avoid algorithm

4.8 Performance Evaluation of the Fourth FishBot Version

Figure 4.15 shows the tracking result of a lure coupled with a FishBot for different speed and acceleration commands. In respect to the existing literature, the FishBot has very high capacities in terms of acceleration (up to 1.3 ms^{-2}), linear speed (up to 30 cms^{-1}), and rotational speed (more than 18 $rads^{-1}$). The maximal speed and acceleration of the fourth version of the FishBot are smaller than the ones of the first version (Sect. 4.3). This is partly due to the increase of weight of the FishBot with the new chassis made of brass to increase its stability, and also the new power management that limits the voltage to 5.5 V for the motors instead of the 6 V required. However, regarding the speed and acceleration of our zebrafish in the constrained environments that will be described in detail in Chap. 8, the FishBot has the required capacities, thus it was qualified for our experiments involving mixed societies of fish and robots.

The last version of the FishBot is also very robust mechanically, thanks to the bearings that were added on the wheels axis. In terms of electronics also, no failures were reported until now, with only the brushes that are subject to wear that need to be manually changed every 50 h of experiments on average.

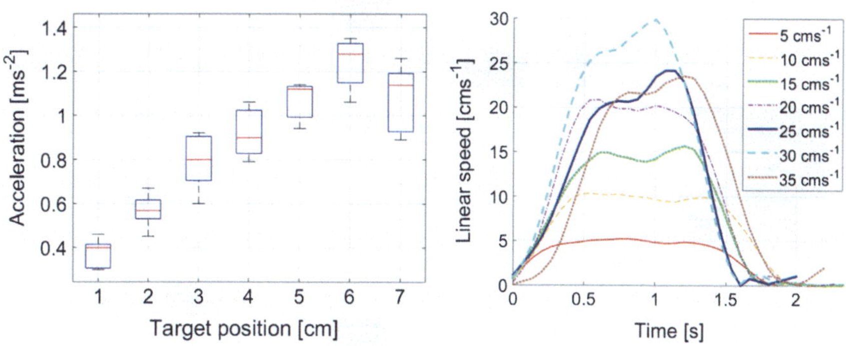

Fig. 4.15 The FishBot speed and acceleration characteristics. Left: The FishBot acceleration measured as a function of the position to reach using a PD position control. Right: The FishBot speed measured as a function of the speed to reach

4.9 Contribution to the State of the Art

With the proposed solution, we can run multi-robot experiments with a high number of robots moving in every possible directions that cannot be achieved by the proposed method in Faria et al. (2010) and Kopman and Porfiri (2013). However, these types of experiment are also possible with the proposed method of Landgraf et al. (2013, 2016) and Swain et al. (2012) that also use a differential drive mobile robot to move a dummy fish underwater through magnetic coupling. When comparing the size of the mobile robots, Landgraf et al. (2013, 2016) use a mobile robot of 70 mm of length and 70 mm of width, while Swain et al. (2012) is using the MiaBot (Corp 2016) that has a length and a width of 75 mm, thus the distance between two fish replicas will be of 70 mm in minimum for the first case and 75 mm for the second case. With the FishBot, thanks to its width of 22 mm, it is possible to have two lures moving at a distance of 22 mm. This is the first main advantage of the FishBot compared to other robots. Moreover, the FishBots are also equipped with IR proximity sensors that can be used to avoid any collisions between the FishBots (Fig. 4.16).

The second advantage is the continuous powering of the system. Indeed, the robotic devices used in Landgraf et al. (2013, 2016) and Swain et al. (2012) are powered using onboard batteries. With our proposed design, the powering of the mobile robot is continuous, which allows experiments that can last up to several days if using a passive lure such as the lures proposed in Landgraf et al. (2013, 2016) and Swain et al. (2012) or the actuated lure RiBot that will be described in the next chapter.

Fig. 4.16 Exploded view of the FishBot parts before assembly that demonstrates the simplicity of the mechanical design of the fourth version

4.10 People Who Contributed to This Work

Dr. Philippe Rétornaz contributed to the electronic and firmware design, Daniel Burnier worked on the design of the first PCB version, Christophe Barraud designed the second version of the FishBot and Norbert Crot worked on the mechanical design, manufacturing and assembly of the three last versions of the FishBot. Alain Berthoud worked on the manufacturing and assembly of the first version. The atelier ATPR of the EPFL manufactured most of the mechanical parts of the final FishBots' production and the atelier ACI manufactured the PCBs. Andr Guignard also offered precious advices during the design phase of all the FishBot versions.

Chapter 5
RiBot, the Actuated Robotic Fish Lure

5.1 Summary

In this chapter, we present the second robotic device that was designed for this project: the RiBot, a robotic fish lure equipped with an actuated caudal peduncle, LEDs, a rechargeable battery, and that can be remotely control underneath the water (Fig. 5.1).

We will first describe the specifications set for the design of this device to perform stimuli analysis with zebrafish and mixed societies experiments using groups of zebrafish and robots, then the different solutions and versions of the robots that were developed throughout this thesis and, finally, the validation of the final version.

This chapter is based on the publication *Infiltrating the Zebrafish Swarm: Design, Implementation and Experimental Tests of a Miniature Robotic Fish Lure for Fish-Robot Interaction Studies* (Bonnet et al. 2016b) and the publication *Design Methods for Miniature Underwater Soft Robots* (Bonnet et al. 2016a) with slight modifications to fit the present dissertation.

5.2 Requirements and General Descriptions

As mentioned in the previous chapter, it was not possible during this thesis to design an autonomous robotic fish lure that could be actuated and at the same time follow a shoal of zebrafish. We therefore designed the FishBot presented in Chap. 4 to steer a lure inside water from underneath the tank through magnetic coupling. However, as presented in Chap. 3, zebrafish, as many other fish, are very sensitive to water vibrations and they probably use their lateral line for sensing during shoaling. Thus, we wanted the robotic device to generate such signals to analyze if this could have indeed an impact on the acceptance of the robot among the group of fish.

F. Bonnet and F. Mondada, *Shoaling with Fish: Using Miniature Robotic Agents to Close the Interaction Loop with Groups of Zebrafish Danio rerio*, Springer Tracts in Advanced Robotics 131, https://doi.org/10.1007/978-3-030-16781-3_5

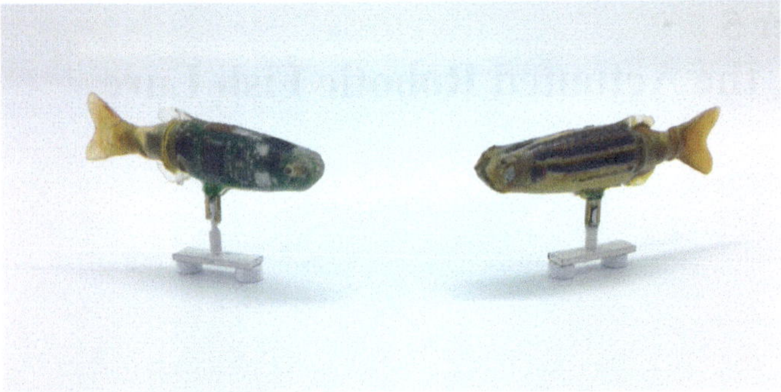

Fig. 5.1 The actuated robotic fish lure RiBot, without and with color pattern respectively

Fig. 5.2 Left: Fishing lure X-Rap Jointed Shad actuated passively using the waterflow (Rapala, Finland). Right: Lure with the same size of the zebrafish that was designed to beat its tail when moving with a certain linear speed. The two parts were designed using 3D printing and painted in order to also mimic the color patterns of the zebrafish

One way to produce vibrations underwater while the lure is moving is to use a passive mechanism that, using the water flow, could move the tail of the lure, as it is done for some fishing lures (Fig. 5.2, left). During this work, we tried to design lures that could at the same time mimic the visual cues of the zebrafish while beating the tail passively (Fig. 5.2, right).

However, we discovered that it was not possible with such design to produce the same types of tail movements as the ones created by zebrafish. Hence, it was necessary to decouple the body movements of the lure from its linear motion underwater. Thus, we opted for the design of a robotic lure equipped with an actuated tail.

All the lures that were designed in this study, such as the one shown in Fig. 5.2, right, were fixed on a module composed of a carbon stick attached to an iron plate on which two magnets were placed. The module was painted in white so that it blends in with the white background of the tank. Hence, the lures moved at a fixed height of 3 cm. In order to increase the visibility of the lure, the water level was always fixed at 6 cm so that the lure was moving at the middle of the water level. This water level was not introducing any stress for the zebrafish as it will be described in Chap. 6.

The design of such underwater robots is always of great challenge, especially at very low scale. The device should be waterproof, wireless and, moreover in the case of this design, the robot has to interact with zebrafish and thus its size shall be in the range of zebrafish, whose average length rarely exceed 40 mm in laboratory conditions. As shown in Abaid et al. (2012a), the zebrafish can be attracted by a bigger replica fish as soon as the latter has the same ratio size as zebrafish. However, preliminary experiments performed using our zebrafish showed that a lure with a size close to the zebrafish was more attractive than a bigger one. Thus, we decided to create a device mimicking the size of zebrafish as much as possible. We named the robotic lure RiBot, which is a combination of the word *Riba* that means fish in Russian language, and the word *Robot*.

5.3 First Version of the RiBot Based on a Rigid PCB

Due to the size of the selected components, the length of the first prototype of the robotic fish lure was fixed at 75 mm while keeping the same ratio size as zebrafish, with a width of 10 mm and a height of 17 mm. For the first prototype, we have decided to include only one actuator, a stepper motor to actuate the tail, a rechargeable battery to allow the energy autonomy of the device and an infrared receiver to remotely control the device underwater.

5.3.1 Actuation Choice

There are several solutions that can be found in the literature to actuate the tail or the fins of a fish-like underwater vehicle tail or fin (Bandyopadhyay 2005). There is also a trend to apply new types of actuators for the development of small-scale biomimetic fish robots (Du et al. 2015). Concerning the RiBot, a very small actuator that consumes very low energy but with enough torque to actuate a robotic fish caudal peduncle underwater was required. The actuator should also allow beating tail frequencies and amplitudes in the range of the zebrafish. Even if this allows high levels of performance to be achieved in terms of size, tail beating frequencies, or energy consumption, the actuators usually suggested in the literature have several drawbacks, such as high-voltage requirements for piezo-electric actuators, temperature sensitivity for shape-memory alloys, specific liquid environment requirements for electroactive polymers, etc., which creates challenges for the design of an autonomous and miniature robotic fish. Moreover, it is sometimes difficult to embed the electronics into the device due to the size of the components required to drive the different actuators; so either the fish need to be rather big, or the electronic need to be partially external, which reduces the autonomy of the system.

There is the solution of standard DC motors coupled with the appropriate transmission using wires or rods to obtain the desired motion. Miniature DC motors can be found on the market nowadays, however, the problem of this solution is the bulky and complex transmission that has to be implemented on a small scale device.

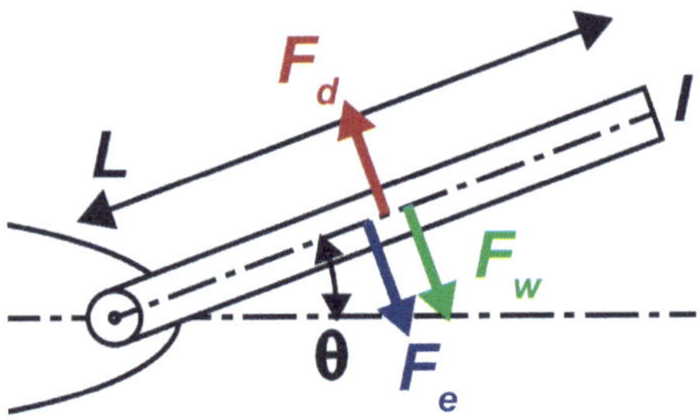

Fig. 5.3 The model of the tail (caudal peduncle) used to estimate the needed torque for the actuator

Another solution is the electro-magnet actuator that is used for instance for the flaps of RC airplanes. These types of actuators are small and have a low power consumption. Furthermore, the amplitude and frequency can be varied. The drawback of this solution is the very low torque of the actuator.

In order to determine the specifications of the motor, we first considered a kinematic fish model shown in Fig. 5.3 to determine the required motor specifications. To simplify the model, we assumed that the caudal peduncle is a rigid body that does circular motion centered on the motor shaft.

If θ is the angle between the tail and the longitudinal axis, the motion equation can be expressed by Eq. 5.1, where I is the inertial moment of the caudal peduncle, L is the length of the caudal peduncle, F_d is the driven force of the motor, F_w is the resistance of the water that is approximated as the drag force only (Eq. 5.2) and F_e is the resistance force of the elastic skin with coefficient k that envelop the caudal peduncle and this resistance will thus be unbalanced on each sides of the caudal peduncle (Eq. 5.3).

$$I\ddot{\theta} = \frac{F_d L}{2} - \frac{F_w L}{2} - \frac{F_e L}{2} \tag{5.1}$$

$$F_w = \frac{1}{2} C_D A \rho \left(\frac{L\dot{\theta}}{2}\right)^2 \tag{5.2}$$

$$F_e = \frac{kL\theta}{2} \tag{5.3}$$

Using Eqs. 5.1–5.3, we obtained a minimal value of 2 mNm for the torque needed for the actuator.

We selected a micro step gear motor MF03G of Seiko Precision Inc. for the actuation of the caudal peduncle, which characteristics can be seen in Table 5.1. The mass of the motor is 0.6 g. The motor has very small dimensions (Fig. 5.4) and can thus be easily integrated inside the robotic lure design. The motor has a maximal torque of 2 mNm and a gear reduction ratio of 1/131 and speed of 1200 step/s. It consumes between 120 and 150 mA. Half step mode can be implemented and offers a fine output precision of 0.171°C/step. The advantage of using a stepper motor is that if the motor does not miss any steps, the position of the caudal peduncle can be estimated from the number of pulses emitted to drive the motor.

Table 5.1 Characteristics of the micro step gear motor MF03G

Characteristics	Value
Motor type	2 phase stepper motor
Size	13.3 × 6.5 × 7.4 mm
Mass	0.6 g
Min. step angle	0.172°
Voltage	3 V
Max. speed	1200 step/s
Internal resistance	28.5 Ω
Torque	Min. 2 mNm

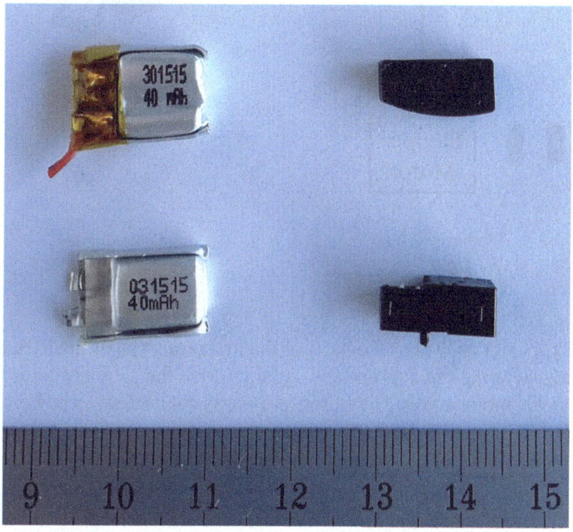

Fig. 5.4 Left: LiPo battery selected for the design. In order to reduce the size of the battery, the charge circuit (in yellow) was removed and replaced by a charge circuit implemented on the Rigid-Flex PCB. Right: The Seiko Precision stepper motor used to actuate the tail of the robotic fish

The original pinion of the stepper motors was removed and replaced by a 3D printed caudal peduncle made of brass for current conductivity as it will be further explained in the next sections.

5.3.2 *Powering*

Figure 5.5 shows the different hardware subsystems of the RiBot. We selected a miniature Lithium Polymer (LiPo) battery (Sparkfun, USA) which can be seen in Fig. 5.4. This battery is rechargeable, thus we do not need to change it on the device when it is empty. Indeed, in our design, as it will described in more detail in Sect. 5.5, the battery can be recharged through the eyes of the RiBot that are made of brass and that cross the external polyurethane coating to be accessed from the outside of the device. The battery protection circuits were removed from the original battery

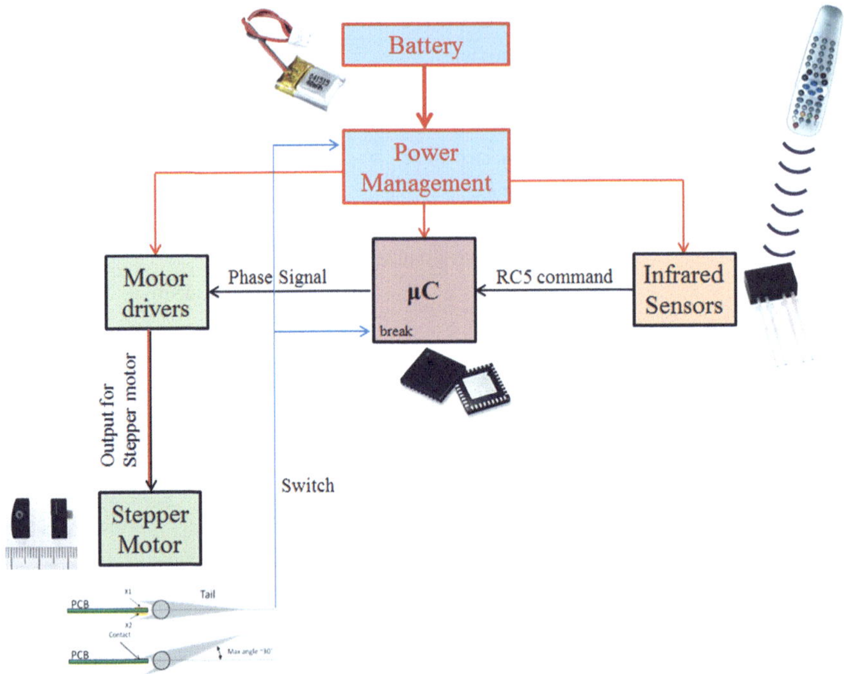

Fig. 5.5 Hardware schematic of the first prototype of the RiBot. The microcontroller STM32f103 is generating the appropriate signal to drive the stepper motor MF03G through a dual full-bridge A3901 motor driver. The battery is directly connected to a power management circuit which manages the charging of the battery and the battery protection in the case of short circuits. The RiBot is turned on when the tail reaches one of the two end position. The RiBot is remotely controlled via RC5 IR signal that is sensed using an infrared sensor TSOP75436WTT

and replaced by a homemade one on the PCB in order to reduce the volume of the battery. A charge circuit was also designed in order to manage the powering of the device and the recharging of the battery.

The caudal peduncle of the RiBot, which is directly connected to the shaft of the stepper motor (see Fig. 5.8), is made of brass and connected to the ground (GND) (Fig. 5.6). It can be used for two purposes: It can turn ON the device, and can also be used to calibrate the tail position. When the caudal peduncle reaches one of its maximal position (Fig. 5.6), an electric contact is made between pads on the PCB and the caudal peduncle. When the device is turned OFF and the tail reaches one of the two pads, the microcontroller is switched ON. From this point, the tail is used for the break function implemented in the microcontroller in order to calibrate the tail initial position. Indeed, there are no sensors placed in order to retrieve the position of the stepper motor, and due to the fact that the stepper motor might lose steps, the tail position has to be recalibrated. The electric contact made between pads on the PCB and the tail is used to determine the two maximal positions of the tail, and thus the zero position located in the middle of these two extreme positions.

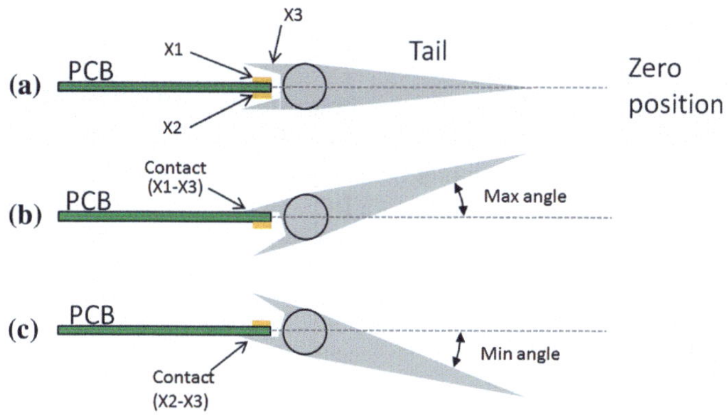

Fig. 5.6 Schematic of the switch system for the tail (viewed from the top of the robot). The actuator is fixed on the PCB and actuates the tail. Two contact pads ($X1$ and $X2$) on the PCB allows an electric contact with the tail when it reaches one of the two maximal position (**b**, **c**). This will either switch ON the device if it was OFF or either generate a break on the microcontroller in order to calibrate the tail position to find the zero position (**a**)

5.3.3 Communication

IR communication is implemented to remotely control the lure. It is a unidirectional communication as no sensing information were needed from the RiBot. An IR sensor TSOP75436WTT is placed on the head of the lure and IR signal can be sent from any direction to control the device underwater. The protocol RC5 is implemented, thus universal TV remote control with RC5 protocol could be used to control the robot.

5.3.4 Electronic Design

The main skeleton of the RiBot's first version consisted of a PCB of 1.6 mm width (Fig. 5.8, left). All the components were soldered on the two faces of the PCB and were encapsulated into an impermeable coating made of polyurethane.

The PCB carried a microcontroller STM32f103 (ST, Switzerland). This microcontroller was selected due to its very small dimensions (6×6 mm), its functionalities and as it was the smallest microcontroller of the ST family.

The stepper motor is driven using a dual full-bridge A3901 also selected for its small dimensions ($3 \times 3 \times 0.75$ mm), its operating range for voltage (2.5–5.5 V) and current ($\pm$400 mA) which made it suitable to drive the actuator

5.3.5 Firmware

The firmware is codded in C language and uses standard STM C libraries to drive the different peripherals of the microcontroller. The code uses two timers, one for the management of the stepper motor control and one for the reception of the RC5 signal. The code is continuously waiting for RC5 commands, and changes the control modes of the motor in function of the received command.

5.3.6 Mechanical Design

The motor with the tail docked on its shaft is fixed inside a part manufactured using 3D printing, called the ring, as it has an elliptical external shape, that is glued on the PCB (Fig. 5.8, left).

In order to isolate the actuator and the caudal peduncle from water and to create a soft skin that can mimic the zebrafishes' tail visual appearance, an undercut of the tail was manufactured using 3D printing. This part was dipped into liquid latex in order to create a thin socket. The socket created was then unmolded and attached to the ring using silicone. A caudal fin, also made of latex, was prepared apart from the tail using another mold, and was glued on the tip of the tail using latex.

Polyurethane is used to isolate the electronics from water. A mold with the desired undercut of the Ribot made of ABS was manufactured using 3D printing (Fig. 5.7). The mold is composed of two parts that are joined using pins and screws during the molding process to press the two parts against each other. The PCB is placed

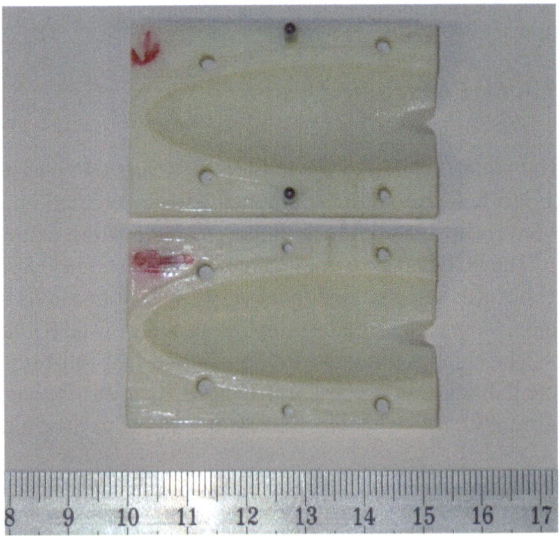

Fig. 5.7 Mold used for molding the first version of the RiBot into polyurethane. The two parts are aligned with pins and then screwed in order to press the two parts of the mold against each other

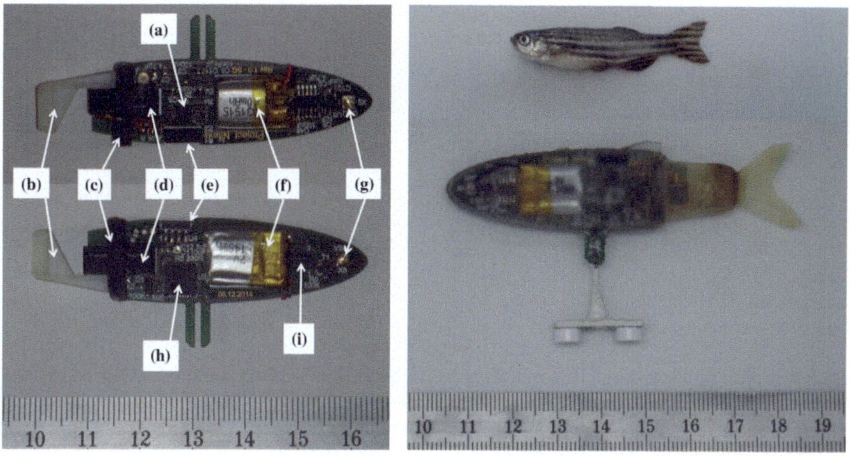

Fig. 5.8 Left: Skeleton of RiBot: **a** Dual H-bridge A3901. **b** Tail (caudal peduncle). **c** Ring. **d** Stepper motor. **e** IR receiver. **f** LiPo battery. **g** Eyes used as contacts to recharge the battery and reference during molding. **h** Microcontroller STM32f103. **i** Connector used for programming the device. Right: First version of the RiBot compared with one of our zebrafish. The device is waterproof thanks to the latex socket (or tail) and the polyurethane coating

inside the mold and the eyes are used as a reference inside the mold. Finally, liquid polyurethane is injected inside the mold from the tail. The polyurethane coats the entire PCB up to the ring and hardens inside the mold. After this process, the RiBot is totally isolated from water and can start to swim underwater (Fig. 5.8, right).

5.3.7 System Qualifications

The first prototype of the RiBot presented in Fig. 5.8 was inserted inside water in order to measure its capabilities.

We could perform 23 min long tests with the tail of the RiBot moving continuously thanks to the embedded LiPo battery. However, when reducing the use of the actuator, the RiBot could be maintained turned ON for more than one hour underwater.

In order to measure the beating tail of the RiBot, we used the same experimental setup that was used to perform the experiments with fish presented in Chap. 6. We installed a color marker on the edge of the tail and above the position of the actuator axis and tracked these markers from the top view using a camera (Fig. 5.9, left). As it can be observed in Fig. 5.9, right, the RiBot has limited amplitude and frequency capabilities that restrict it from reproducing some of the extreme body movements of zebrafish, which can bend their caudal peduncle with an angle bigger than $\pi/2$ rad (90°) with an angular speed over 4 rads^{-1}. The maximal beating amplitude of the RiBot is 0.52 rad (30°), and the maximal angular speed is 3.6 rads^{-1} but for very low amplitudes. However, the RiBot is able to reproduce the average turn rate of the zebrafish ($\sim$2.8 rads^{-1} (Mwaffo et al. 2014)) for an amplitude of 0.32 rad (18.35°).

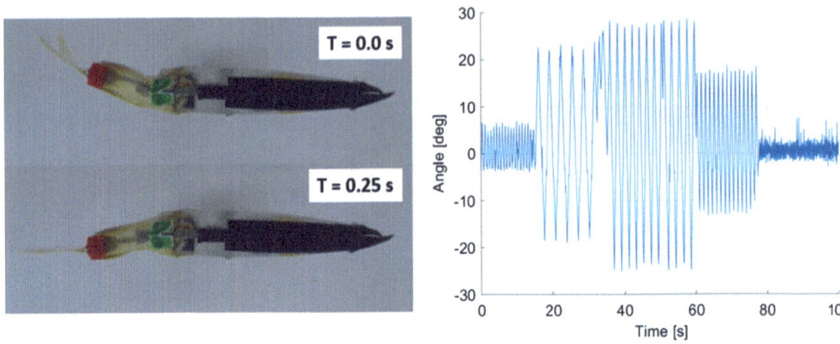

Fig. 5.9 Left: Top view of the RiBot beating its caudal peduncle. Two markers are placed to track the movement of the tail. Right: Results of the tracking of the caudal fin for different tail beating amplitudes and frequencies. The frequencies and amplitudes of the modes are the following: 5 Hz, 5°; 1 Hz, 20°; 1 Hz, 30°; 2 Hz, 15°; 10 Hz, 2°

5.4 Second Version Based on a Rigid-Flex PCB

Based on the lessons learned from the first design, the preliminary experiments obtained with zebrafish and the known zebrafish stimuli perception, we draw up the following specifications for a second version of the RiBot: The dimensions of the lure needed to be reduced and as close as possible as the one of a real zebrafish. It should provide more biomimetic aspects, such as more realistic fins, and body features. The robot had to integrate RGB LEDs in order to generate a new type of stimuli never tested on zebrafish with such robotic lures, as only different color surfaces were tested in published studies. Also, the LEDs could be used for debug purposes. Finally, the device should be reprogrammable once molded which was not the case for the first design.

5.4.1 Rigid-Flex PCB Design

In the first RiBot design, a two copper layers PCB was used as a skeleton (Fig. 5.8). This PCB offered enough room to mount the electronic components and routing. However, with the new specifications, it was not possible to use such a simple design to integrate the new components while decreasing the size of the lure. The solution was the use of a Rigid-Flex PCB.

Rigid-flex PCB is the name given to a PCB that is a combination of both flexible and rigid circuits. Some parts of the PCB can be composed of only flexible materials that can be bent and act as interconnections between the rigid sections. Some parts can be flexible but with a solder mask in order to sold some components more easily. Finally, some parts can be rigid to carry all the bulky components and provide a rigid mechanical reference. The number of copper layers can also be varied. Rigid-Flex

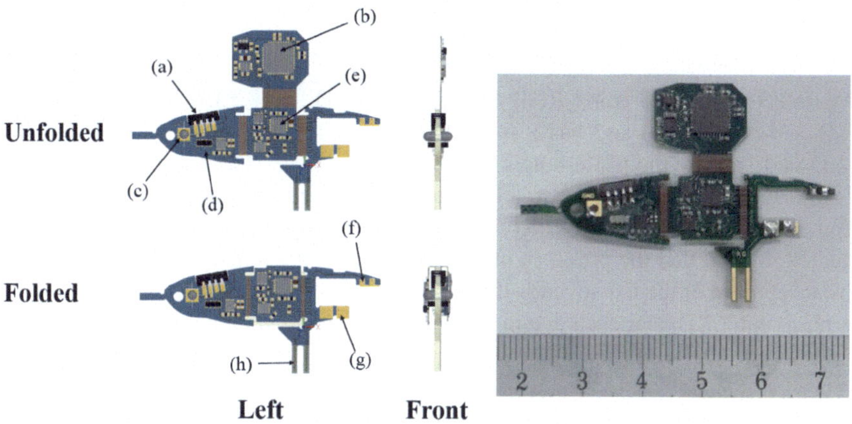

Fig. 5.10 Left: The Rigid-flex PCB design of the RiBot V2.0 unfolded (top) and folded (bottom). The flexible parts are folded with two 90° angle folds. The blue colored areas indicate the areas with solder mask, while the brown colored indicate the flexible parts without electronic components. Right: Prototype of the Rigid-Flex PCB with the electronic components soldered. The head and the tail parts are rigid with two layers of solder mask (green). The left and right fillets are flexible with one layer of solder mask on which the components are soldered. The flexible joints between the parts that contain electronic circuits are colored in brown due to the color of the cover layer as they are not composed of a solder mask

PCBs thus offer the ability to design the circuitry to fit in a more optimal way a three dimensional device. Indeed, the bulky connectors that are usually used to transfer the information between two PCBs are replaced by flexible surfaces that are parts of the whole PCB and that can bend.

Figure 5.10 shows the design of the Rigid-Flex PCB realized for the second version of the RiBot. It is composed of four parts that are separated by flexible joints that only contains circuit tracks:

- The head which is rigid and composed of four copper layers. This part contains the eyes that are used to recharge the device (Fig. 5.10c), the RGB LEDs and their drivers (Fig. 5.10d), the IR receiver (Fig. 5.10a) and the battery protection circuits.
- The left fillet which is flexible, composed of two copper layers and a solder mask layer. This part contains the motor drivers (Fig. 5.10e) and the battery charge circuits.
- The right fillet which is flexible, composed of two copper layers and a solder mask layer. This part contains the microcontroller (Fig. 5.10b), a voltage sensor to measure the battery voltage and a voltage regulator.
- The tail which is rigid and composed of four copper layers. This part is used to solder the stepper motor (Fig. 5.10f), connect the programmer (Fig. 5.10h) and create a contact between the caudal peduncle and the PCB in order to switch ON the device or detect the end position of the caudal peduncle (Fig. 5.10g).

The Rigid parts are 1.2 mm thick and the flexible parts 0.22 mm. The choice of 4 copper layers for the rigid parts was mandatory for routing purposes.

5.4.2 LED

We selected the full-color RGB OVSRRGB LED. This LED is very compact ($3.2 \times 1.5 \times 1$ mm) and has a maximum luminosity intensity of 450 mcd. The LED is driven by the triple output I2C chip NCP5623C. The I2C protocol allows us to drive the two LEDs separately using only two circuit tracks (SCL and SDA) from the microcontrollers to the two chips that have different I2C addresses. Some studies showed that the presence of realistic eyes on the lure design could increase the acceptance of the fish towards the device (Landgraf et al. 2016), thus we decided to place one LED near each eye in order to increase the visibility of this part of the body for the zebrafish.

5.4.3 Mechanical Design

The external shape of the second version of the RiBot was defined using the 3D scan of a dead zebrafish (Fig. 5.11). The zebrafish was scanned using a Stereoscan 3D Breuckmann StereoSCAN3D with two cameras of 1.4 megapixels. The surfaces retrieved by the scanner were processed and scaled in order to design a mold with the desired undercut of the zebrafish in which the Rigid-Flex PCB could fit. The mold was made of ABS using 3D printing as it was done for the first version of the RiBot.

The fins of the fish were made using Mylar™ as it offers good softness and is available in very thin films. The fins have a thickness of 50 μm and were also designed from the 3D scanned of the fish.

5.4.4 Assembly

The electronic components were soldered prior to the folding of the PCBs in order to facilitate their positioning. A solder stencil was designed to deposit the solder material, and then the PCB was put into a vapor phase oven to solder the electronic components. The eyes, the stepper motors and the battery were soldered on a second phase using a soldering iron (Fig. 5.12).

Fig. 5.11 3D scan of a male zebrafish that was obtained using a Stereoscan 3D Breuckmann StereoSCAN3D with two cameras of 1.4 megapixels

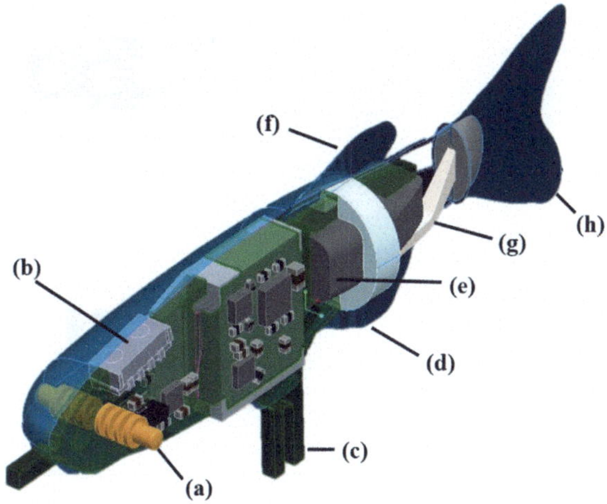

Fig. 5.12 The mechanical design of the lure was made using PTC CREO Parametric, educational edition. This software allows surfaces design that could match the 3D scan of the zebrafish. It was used to design the Rigid-Flex PCB shape, the caudal peduncle, the fins and the molds. **a** Eyes. **b** IR sensor. **c** Contacts to reprogram the device and attached the lure to a module equipped with magnets to be coupled with the FishBot. **d** Anal fin. **e** Stepper motor. **f** Dorsal fin. **g** Caudal peduncle. **h** Caudal fin

Figure 5.13 shows the procedure to fold the PCB. First the PCB is mounted on a support and two metallic wedges are screwed to constrain the PCB and mark the fold (Fig. 5.13a). The two flexible parts between the left fillet, the head and the tail are folded with two 90° angles (Fig. 5.13b). This process is very delicate as only one fold can be done, otherwise the flexible part would break, and therefore it is irreversible. The battery is then inserted and glued on the left fillet (Fig. 5.13c). Finally, the right fillet is folded around the battery and glued on it (Fig. 5.13d).

The assembly steps of the different components of the RiBot, as well as the molding steps can be seen in Fig. 5.14. In order to isolate the actuator and the caudal peduncle from water and to create a soft skin that can mimic the tail of the zebrafish, an undercut of the tail was made in 3D printing as it was done for the first version of the RiBot. The skin created was unmolded and attached on the ring using silicone. The caudal fin, the anal fin and the dorsal fin are made using thin films of Mylar™. The Mylar™ is cut using laser cutting and then glued on the tail using latex. The, the PCB is placed in a mold and polyurethane is injected following the same process of the first version of the RiBot.

5.4.5 Performances Evaluation

The second version of the RiBot (V2.0) is presented in Fig. 5.15. We are satisfied of the results obtained in terms of biomimetics appearance and size ratio. The last aspect

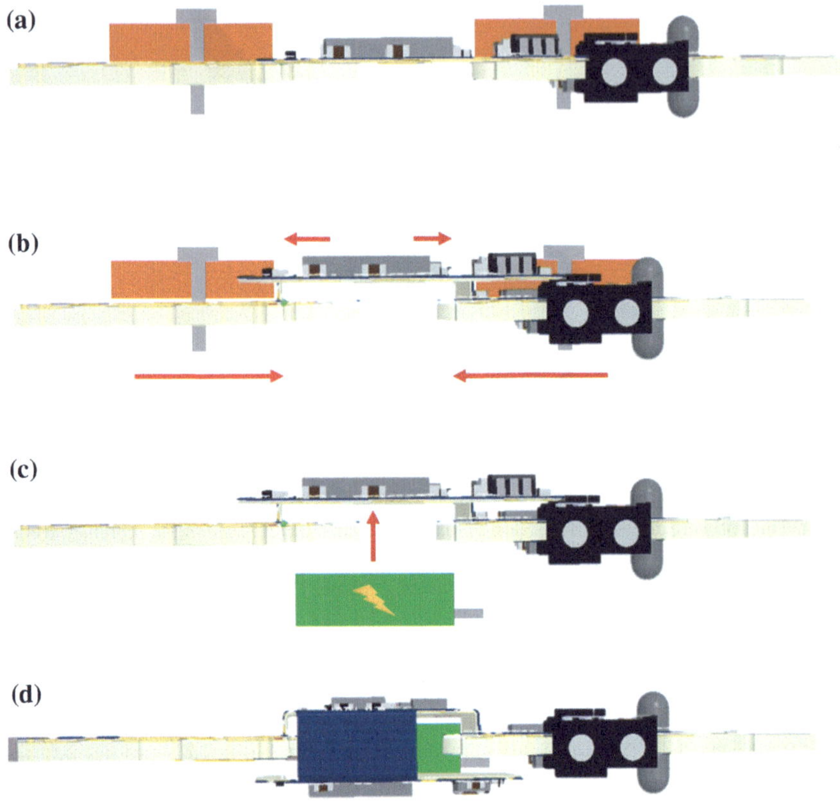

Fig. 5.13 Folding process of the PCB viewed from the top. **a** Some metallic wedges are fixed on the PCB. **b** The PCB is folded against the wedges with two 90° angles. **c** The battery is placed and soldered onto the PCB. **d** The right fillet is folded around the battery and glued on the battery

that still need to be considered is the coloration of the lure. Some lures RiBot were painted in order to analyze if the visual appearance has an effect on the zebrafish, and results of this analysis can be seen in Chap. 10.

Concerning the technical aspect of the prototype, we performed a successful remote controlled experiment with the lure below 30 cm water layer. The waterproofness of the lure was validated. The actuated tail was found to run well, with amplitude measured between 0 and 23° and frequencies between 0 and 20 Hz which are similar to the first version of the RiBot. We could perform 23 min long test with the tail of the lure moving continuously using the embedded LiPo battery. The LED could also be well seen through the polyurethane layer (Fig. 5.16).

During the folding process, we encountered some issues on the components that were soldered near the flexible part (Fig. 5.17). On the design, some electronic components were located very close to the flexible part in order to use all the available surface, thus the distance between the contact pads on which the components are

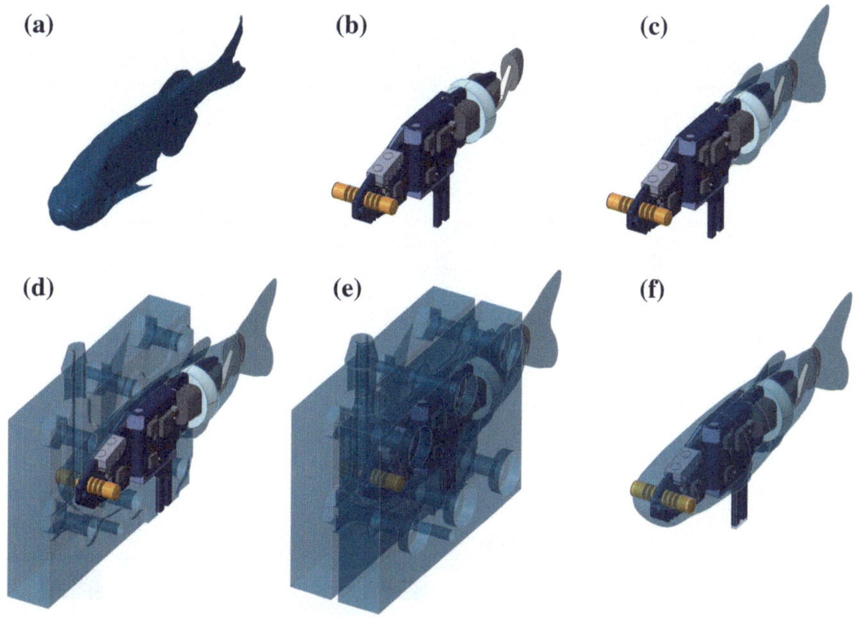

Fig. 5.14 Assembly steps of the RiBot V2. **a** The scan of the zebrafish that is scaled up 1.5 times. **b** The RiBot with all the electronic components, the ring, the motor and the caudal peduncle mounted. **c** The RiBot with the tail, caudal, dorsal and anal fins. The tail is glued on the ring using silicone, and the fins are glued on the tail using latex **d** The RiBot placed inside one of the two molds part. **e** The RiBot is inside the two molds, and the polyurethane is injected from the tube in the front side of the mold. **f** The RiBot ready to swim

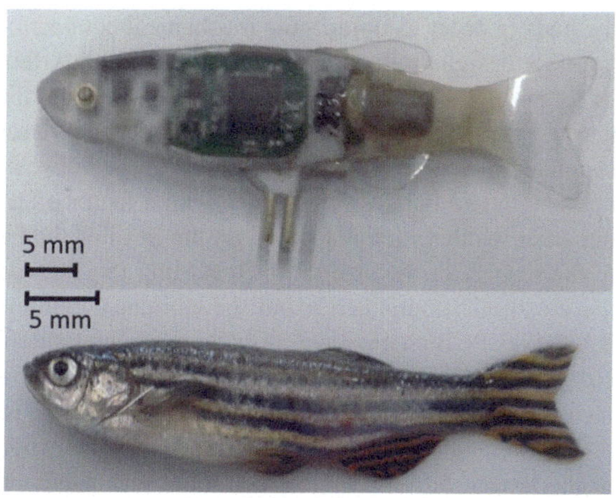

Fig. 5.15 Second version of the RiBot fish lure compared with one of our zebrafish. The latex tail is mounted with the Mylar™ fins. The front part of the lure is molded into polyurethane and only the eyes made of brass are crossing the polyurethane layer. The eyes are used to recharge the device and the four contacts on the pelvic fin are used to reprogram and debug the device

soldered and the flexible part were less than 1 mm. Once folded, the constraints on the electronic circuit in the rigid part were thus too big for the components located very close and thus we encountered cracks either between the pads and the copper

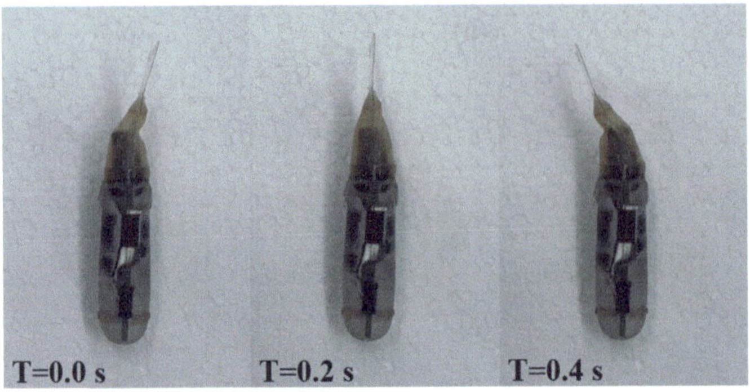

Fig. 5.16 Top view of the lure with its tail beating with an amplitude of 23° and a frequency of 1.25 Hz

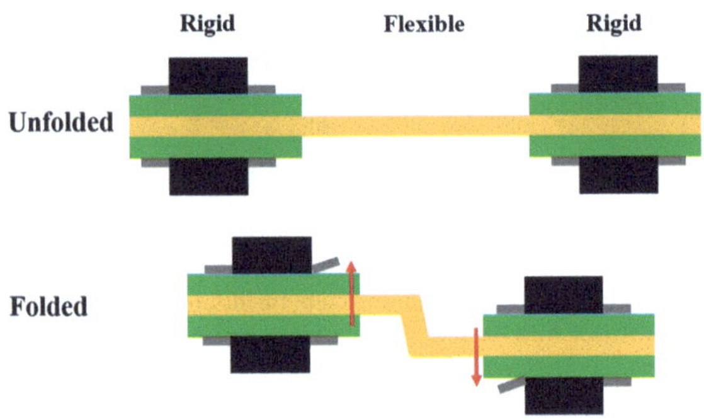

Fig. 5.17 Top: Rigid-Flex PCB unfolded. The components are soldered on both side of the rigid parts. Bottom: Rigid-Flex PCB folded with the flexible parts bent two times with 90° angles. The pads of the components very close to the flexible parts may break during the folding process

layer either between the pin and the pad. The solution for this design was to glue the pads in order to rigidify the mechanical connection. This could be improved on a further design, with components located slightly farther than the flexible part.

Regarding the improvements on the new design of the RiBot lure compared with the first one presented early in this chapter, we describe in Table 5.2 some of the characteristics to demonstrate the differences between the two versions. The maximal length and height were reduce with a factor of 1.26 and 1.42 respectively. The maximal width which is located on the head of the lure was slightly increased due to the respect of the size ratio. Regarding the PCB surface available for soldering the electronic components and the estimated volume, we see that, thanks to the Rigid-Flex PCB technic, we could increase the available surface to mount electronic components with a factor of 1.16 while decreasing the volume of the system with a factor of 1.56. The mass of the new design was also reduce with a factor of 1.77.

In general, the use of Rigid-Flex PCB was really beneficial for the miniaturization of the lure. It allowed us to place more electronic components, reduce the size of the device and mimic most of the body features of a zebrafish. We think that this methodology of using Rigid-Flex PCB methods to fit electronics components into a biomimetics design will help scientists during the miniaturization step of their devices for biomedical robotics and biomechatronics applications.

Table 5.2 Comparison of some characteristics between the first and second design of the RiBot

Version	1	2
Max. length [mm]	80	63
Max. width [mm]	10	11
Max. height [mm]	20	14
PCB surface [mm^2]	858	1002
Estimated volume [mm^3]	8320	5390
Mass [g]	10.48	5.9

5.5 RiBot Battery Charger

We designed a charger specifically dedicated to the RiBot (Fig. 5.18). We used the miniature LiPo single cell charger MCP73831 (Microchip, USA) to manage the charging of the LiPo battery. The device consists of two contact springs that are fixed on a PCB that integrates the charger chip, and are applied on the eyes of the RiBot to guarantee the contact. The charger can be connected via micro-USB cable to start the charging of the RiBot. A LED indicates the charging status.

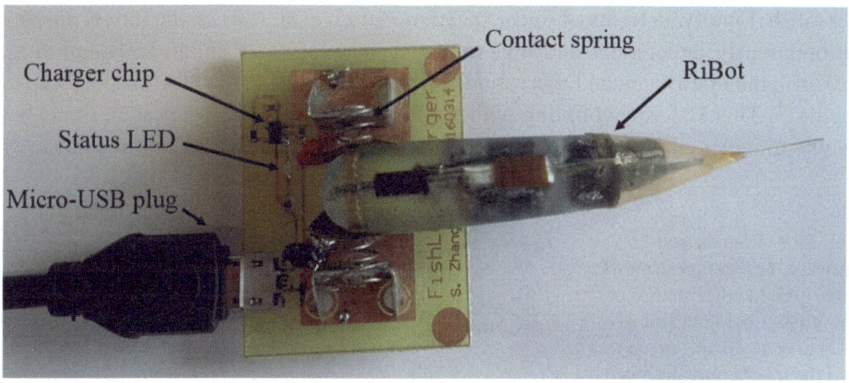

Fig. 5.18 RiBot charger. The charger can be connected via micro-USB to charge a RiBot. It takes maximum 50 min to fully recharge a RiBot

5.6 Contribution to the State of the Art

5.6.1 *Robotic Fish Lures for Animal-Robot Interaction Studies*

Figure 5.19 shows the comparison between the external shape of different designs used in similar research of interaction between zebrafish *Danio Rerio* and robotic lures of Abaid et al. (2012a), and the RiBot device designed in this study. We can see that the RiBot is almost two times smaller in length than the lure presented in Abaid et al. (2012a). Moreover, in terms of size ratio, Fig. 5.15 demonstrates that the RiBot was designed respecting the size ratio of a zebrafish as it was based on a 3D scan of a zebrafish that was scaled so that the electronic and mechanical components could fit into, which was not the case in our opinion with the lure used in other similar studies. In terms of tail beating frequency, the robot that is used in most of other studies involving zebrafish presented in Kopman and Porfiri (2013) has less capacities than the RiBot which can reach 20 Hz of beating tail frequency.

Table 5.3 shows a comparison of other actuated robotic lures designed for other studies on fish-robot interaction with the RiBot and a zebrafish. In terms of size, we have smaller length, width and height while respecting more the shape ratio of a zebrafish. Finally, in terms of linear speed, in Aureli et al. (2012), the lure is moving autonomously underwater and thus its linear speed is quite small. In Butail et al. (2014b), the lure is attached to a robotic arm that allows it to move with speeds up to 4 cms^{-1}. Thanks to the coupling with the FishBot wheeled mobile robot, the Ribot is able to move with much higher speeds than the other published solutions.

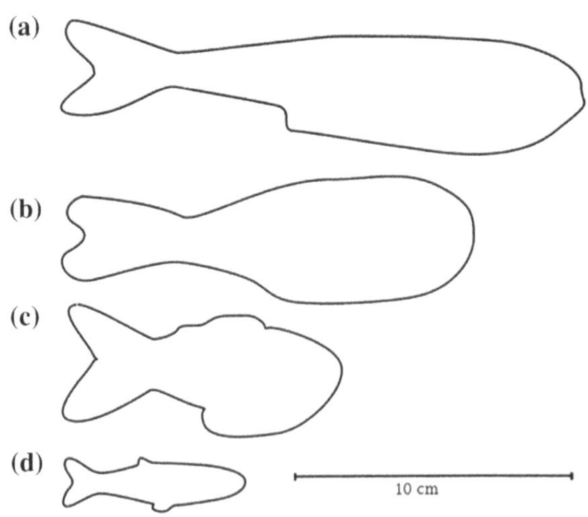

Fig. 5.19 Comparison between the external shape of lures used in other studies of zebrafish-robot interactions (Abaid et al. 2012a) (**a**)–(**c**), and the robotic lure designed in this study **d**

Table 5.3 Comparison of the size and the linear speed of the robotic lures presented in other similar studies on fish-robot interaction

System	Length [mm]	Height [mm]	Width [mm]	Speed [mms^{-1}]
Abaid,2012	150	48	26	Not moving
Aureli, 2012	90	45	35	8–12
Butail, 2014	117	48	26	40
Ribot	63	15	11	300
Zebrafish	45	10	5	450

5.6.2 Bio-Inspired Miniature Robotic Fish

The RiBot is also the smallest bio-inspired robotic fish that can be found in the literature. Of course, compared to other robotic fish used for research in underwater fish-like locomotion for instance, the RiBot cannot achieve autonomous swimming. However, it is able to propel itself in water using the actuated tail. The one degree of freedom actuator, which actuates the caudal peduncle, coupled with the thin caudal fin, allow the device to move forward autonomously underwater with speeds of up to 2.5 cms^{-1}, using a floating element to stabilize it as no elements to control the buoyancy are implemented yet. This maximal speed is in the range of some results obtained with micro underwater vehicles (Wang et al. 2008; Heo et al. 2007).

We measured the linear speed of RiBot underwater with all the different possible amplitudes and frequencies. It can be observes in Fig. 5.20 that RiBot swims slightly faster at high amplitude than at high frequency. The maximal speed of 2.5 cms^{-1} is obtained at an amplitude of 22° and a beating rate of 1 Hz.

We also investigated the possibility to add a second actuator on the RiBot, so that it could swim autonomously underwater as it is done for instance for the Jessiko Robot (Jes 2016). But we would require probably a third version of the design to integrate it, and the robot could not be used as a tool for fish-robot interaction experiments due to the fact that it would embed non-biomimetic features and would still not be able to reach the required speed and acceleration that are provided by the FishBot. However, using the current technology, we think that is possible to create such miniature autonomous mobile robotic fish.

5.7 People Who Contributed to This Work

Daniel Burnier and Norbert Crot are the main contributors of the RiBot design as they worked respectively on the electronic and mechanical designs and made a tremendous job. Samuel Goy and Yuta Kato also contributed for the design of the first version of the RiBot. Shujie Zhang made the prototype of the charger. We also thank Marion Segall, Dr. Anthony Herrel and Dr. Ramiro Godoy-Diana for the 3D scan of the

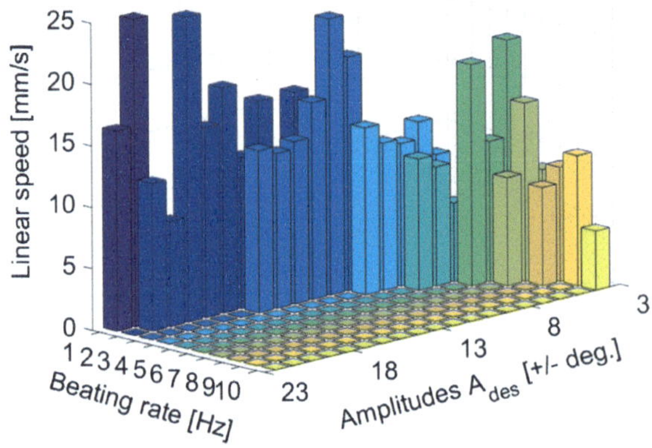

Fig. 5.20 Average linear speed of RiBot underwater in function of the amplitude and the beating rate of the tail. We performed 10 measurements for each combination. The RiBot was attached to a floating mass for stabilization during the speed measurement. The value of 0 corresponds to the cases beyond the motor capabilities and thus impossible

zebrafish that was used to make the design of the mold of the second version of RiBot. Finally, we would like to thank TOM-IC company for producing the Rigid-Flex PCBs.

Chapter 6
Automated Setup to Conduct Experiments with Mixed Societies of Fish and Robots

6.1 Summary

In this chapter, we provide an overview of the experimental setup that was designed for mixed societies experiments with zebrafish and the two robotic devices described in the two previous chapters (Fig. 6.1). This setup design was implemented three times. One time at the EPFL in order to perform the validation tests and preliminary biological experiments, as well as two times at the University Paris Diderot in Paris to run the necessary biological experiments for the ASSISIbf project.

6.2 Experimental Setup

To monitor the behavior of the zebrafish and control the robots in a closed-loop, a dedicated experimental setup was needed. Indeed, all the tests made involving the zebrafish should be performed in the same environmental conditions for the robustness of the results. A clean environment with controlled water conditions including tools to retrieve high-resolution images were needed for this study. Also, the robots, due to their specific designs, required a dedicated infrastructure for the powering and their control in a closed-loop.

6.2.1 *The Experimental Tank and Surroundings*

Experiments were carried out in a $100 \times 100 \times 25$ cm experimental tank made of glass with its internal walls covered with white adhesive (Fig. 6.2). In addition, the bottom surface of the tank was covered on the inside with white teflon plates to avoid the reflection of images on the glass and to obtain a smooth surface for the

F. Bonnet and F. Mondada, *Shoaling with Fish: Using Miniature Robotic Agents to Close the Interaction Loop with Groups of Zebrafish Danio rerio*, Springer Tracts in Advanced Robotics 131, https://doi.org/10.1007/978-3-030-16781-3_6

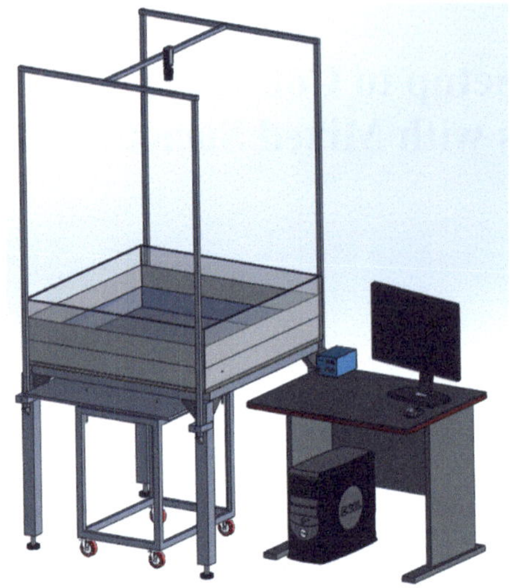

Fig. 6.1 Automated experimental setup designed to conduct experiments involving mixed societies of fish and robots

motion of the lure modules inside the aquarium. The tank is placed on a supportive structure made of aluminum, with four pillars connected by horizontal bars. This structure offers a sufficient space below the tank to place the conductive table on which the FishBots are moving. The tank is filled with water up to a level of 6 cm with temperature set to 27 °C. This level of water is not introducing more stress for the fish (Reed and Jennings 2011) and, furthermore, the lure modules, whose height cannot vary, will be more visible for the fish that are swimming around. The upper part of the setup is confined behind white sheets to isolate experiments from the rest of the room and to homogenize the luminosity (Fig. 6.4e). The bottom part of the setup, i.e., below the aquarium, was isolated from the ambient light using black sheets (Fig. 6.4g), in order to increase the efficiency of the visual tracking used to track the FishBots from below as it will be described in Chap. 7. The FishBots are moving underneath the aquarium, and the motion is transmitted to the lure modules using magnets (Fig. 6.3).

6.2.2 Cameras

We used an overhead acA2040-25gm monochrome GigE CCD camera (Basler AG, Germany) with a maximum resolution of 2048 × 2048 px and equipped with low distortion lenses CF12.5HA-1 (Fujinon, Tokyo, Japan) to grab high resolution frame that were processed on a computer.

To track the FishBots from below (see Chap. 7), we selected a fisheye lens full HD 1080P USB camera module USB2.0 with OV2710 Color sensor MJPEG with

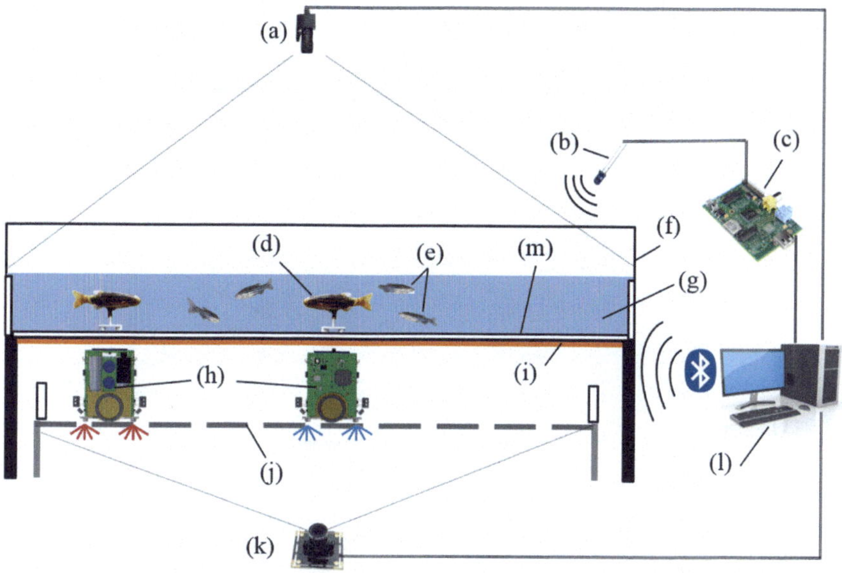

Fig. 6.2 Schematic of the experimental setup. **a** Basler camera used to grab high-resolution frames to track the lures and the zebrafish. **b** IR emitter to emit the controlled commands for the RiBot. **c** Raspberry PI to generate the required controlled commands to send to the RiBot. **d** Fish-lure RiBot inside the aquarium linked to the mobile robot through magnetic coupling. **e** Zebrafish. **f** Aquarium of 100 × 100 × 25 cm. **g** Water of 6 cm depth. **h** FishBot mobile robot moving under the aquarium. **i** Copper conductive plates to power the mobile robot (VCC). **j** Perforated stainless steel plates to serve as ground contact for the FishBot (GND) and to observe the FishBot LEDs from below. **k** 180° fisheye camera to track the FishBot from below. **l** The control station that runs CATS tracking and control software (see Chap. 7). **m** Teflon plate covering the bottom of the experimental tank

a 180° lens from the company (ELPCCTV, Guangdong, China). We were forced to use a wide-angle vision camera due to the small distance between the tank and the ground.

6.2.3 *Computer*

In order to run the control and tracking software CATS that will be described in the next chapter, and at the same time stream and record the high definition videos, a powerful workstation was required. A Dell Precision T7910 with a Double processor Intel® Xeon® E5-2630 v3 was selected. The workstation was running an Ubuntu Linux 14.04 operating system. The graphic card selected was a NVIDIA NVS 315 of 1 GB and the memory card is a 32G 2133 MHz DDR4 (8×4 GB) RDIMM ECC. Due to its high capabilities, this workstation was also used to per-

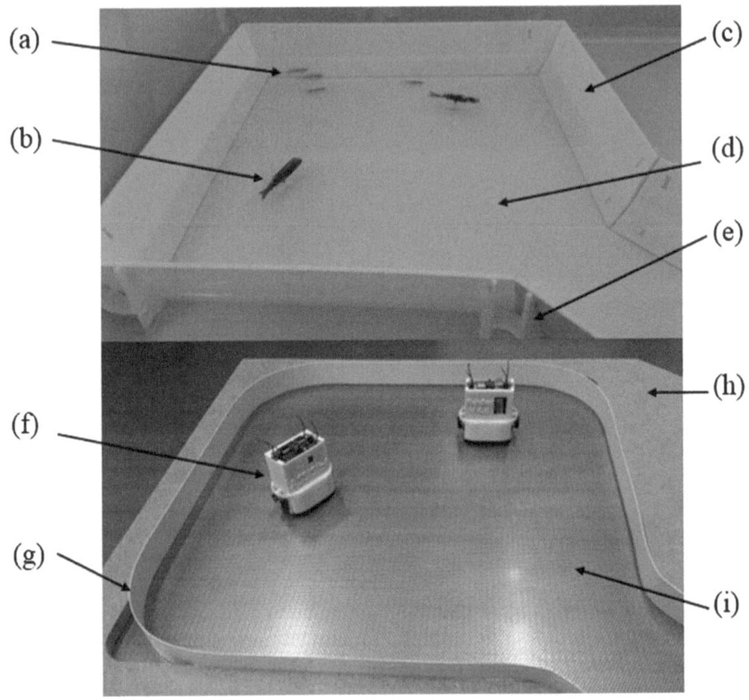

Fig. 6.3 View inside the top and bottom parts of the experimental setup. **a** Zebrafish. **b** Fish-lure RiBot. **c** Wall of the arena. **d** Aquarium floor made of teflon sheet. **e** Pillar of the arena. **f** FishBot mobile robot. **g** Wall of the FishBot arena to constrain de FishBots. **h** FishBot arena. **i** Stainless steel grid

form the post analysis to retrieve the identification of the zebrafish that required high processing.

6.2.4 Lightening

The experiments were performed under daylight conditions. The lightening was done using six lamps Osram FQ HO of 49 W, 90–100 Ra and 6500 K to reproduce daylight conditions. The lamps were disposed alongside the experimental setup, and spotted the light in the direction of the aquarium. The structure on which the tank is fixed was covered with white sheets in order to diffuse the light inside the tank.

6.2.5 Continuous Powering System of the FishBot

The FishBot mobile robots are powered by two conductive plates, one glued onto the bottom of the aquarium and one onto a plexiglass plate on which the FishBots are

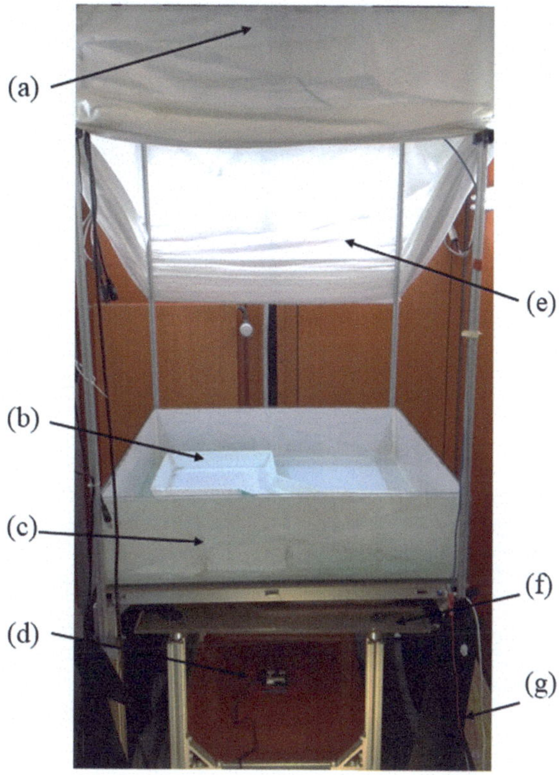

Fig. 6.4 View of the experimental setup implemented at the EPFL. **a** Position of the high-resolution camera. **b** Fish arena. **c** Tank. **d** Bottom fisheye camera. **e** White sheets that diffuse the light. **f** Structure to support the FishBots. **g** Black sheets to isolate the bottom of the aquarium from ambient lightning

moving. The latter is made of stainless steel of $100 \times 100 \times 0.1$ cm perforated with 1 mm diameter holes that are distant of 2 mm and disposed in a triangular shape. This allows the tracking of the LEDs of FishBots with the 180° fisheye camera installed under the setup while still being able to continuously power the FishBots, as it will be further explained in detail in Chap. 7.

6.3 Arenas

We designed several types of arenas to constrain the zebrafish movements inside the tank (Fig. 6.5). These arenas were designed based on two main factors: first, the behavior of the zebrafish, i.e., movements and shoaling, can vary depending on the shape of the environment as shown in Sect. 3.6, and, second, the shape of the setup can improve the measurements of the collective decisions of the zebrafish, for instance using a binary choice setup. The different tests that were performed with mixed societies of fish and robots that are described in the next chapters of this dissertation will show in more detail how the shape of the arenas influence the fish

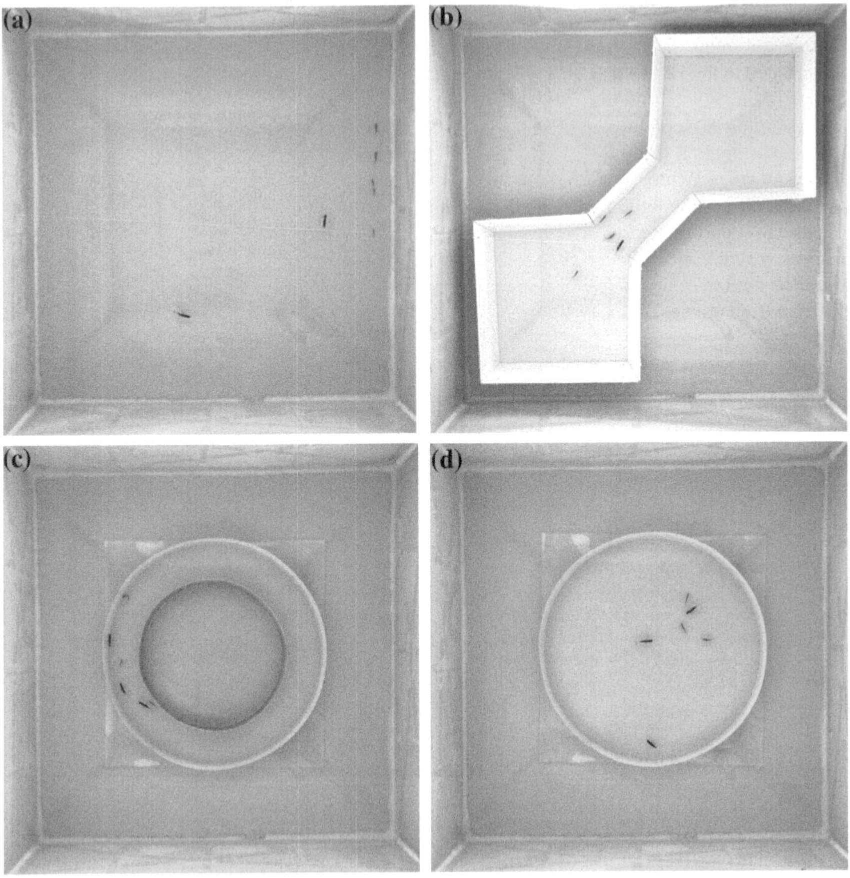

Fig. 6.5 The arenas that were used to perform behavioral experiment with zebrafish. **a** The open area of dimensions 1 × 1 m. **b** The two rooms setups, with two rooms of 35 × 35 cm linked by a corridor of 10 cm width. **c** Circular corridor arena. The width of the corridor is 10 cm. **d** The open circular arena. The external diameter of the circle is 58 cm

behavior and may facilitate the measurements of the fish behaviors. To also constrain the movements of the lures in the arena inside the tank, an arena was also designed to constrain the movements of the FishBots underneath the tank (Fig. 6.3h). This arena is made of wood that is laser cutted in order to have the same shape of the arena inside the tank. It is also covered with white adhesive on this inside for a better reflection of the IR signals, in order to increase the efficiency of the local obstacle avoidance.

6.3.1 Open Arena

The open area (Fig. 6.5a) consists of a square without any obstacle inside. The square has a dimension of 1 × 1 m. In this setup configuration, the zebrafish have a tendency to follow the walls. We used this configuration to perform behavioral experiments in open area (see Chaps. 8 and 9).

6.3.2 *Two Rooms with a Corridor*

This setup was designed to offer a binary choice for the zebrafish (Fig. 6.5b). In fact, the zebrafish can either be in one of the two rooms, either in the corridor. However, they will spend much of the time inside the two rooms, transiting rapidly by the corridor. The size of the corridor was designed so that the zebrafish would not spend too much time in it and avoid as much as possible to make U-turns. The width of the corridor is 10 cm and the two rooms have a dimension of 35 $\times$ 35 mm. The walls of this arena are slightly inclined towards the inside, so that no occlusion are made due to the walls, and all the agents are always in the field of view of the top camera. The walls of the arena are made of Polymethyl methacrylate (PMMA).

This setup was designed mainly for the experiments that took placed in Paris and that were conducted by the biologists involved in the ASSISIbf project. Some of these experiments, especially the ones involving the robots are still in progress and thus, they will not be described in this thesis.

6.3.3 *Circular Corridor*

As the zebrafish are constantly moving, we proposed the use of a circular corridor to measure the attraction of the lures (Fig. 6.5c). This setup offers a binary choice for the fish, as they can either move in the clockwise direction either in the counter clockwise direction. It is also a known setup to study the shoaling formation of zebrafish (Abaid and Porfiri 2010).

The dimensions of the corridor are the following: an external diameter of 58 cm, an internal diameter 38 cm, thus the width of the corridor is 10 cm. The choice of the 10 cm width is a good tradeoff to have a continuous motion of zebrafish without stressing them due to the lack of room. A more detailed description of this setup and its use will be done in Chaps. 10 and 11.

6.3.4 *Open Circular Arena*

This setup was designed to constrain the zebrafish into a smaller room that in the open area, without corners in order to avoid the aggregation of the zebrafish in one place (Fig. 6.5d). In this setup indeed, compared with the open area, the linear speed of the zebrafish will be more constant. The external diameter of the circle is 58 cm.

In this configuration, the zebrafish have a tendency to move in shoal and occupy the entire arena (see Sect. 8.5).

6.4 Experimental Procedure

Here, we summarize briefly the experimental procedure that was used for each experiment involving zebrafish that are described in this dissertation. The experimental procedure was validated by the Department of Consumer and Veterinary of Canton de Vaud (Switzerland).

First, the water of the experimental tank was maintained at the same temperature (27 °C) and water quality as the water of the housing aquarium (see Sect. 3.7) to minimize the effect of the water transition on the zebrafish. In the morning of an experiment session, zebrafish were selected at random from their housing aquarium and were maintained inside a transfer tank next to the experimental tank during the experiment. Then, a small group of zebrafish was selected among the entire group with a hand net from the transfer tank and transferred into the experimental tank. We let them acclimatize for about 10 min inside the experimental tank before starting an experiment, as we noticed that for the first five to ten minutes, the behavior of the zebrafish is not the same than the behavior during the rest of the experiment, probably due to the high-level of stress that is due to the transfer and acclimatization to the new environment. After the experimentation, the fish were placed in a second transfer tank near the experimental setup, so that they could not be reused during the same day for an experiment. After the experiment session, all the fish were put back into their housing aquarium.

Chapter 7
CATS, the Control and Tracking Software

7.1 Summary

This chapter introduces the high-level software infrastructure developed for automated experimentation and analysis of mixed groups composed of robots and fish (Fig. 7.1). We will first describe the general architecture of the software. Then, in more detail, we describe the tracking tools that were used to retrieve the position of the robots and the zebrafish during the experiments, as well as the interface to control the robots. The different implementations of robot's behaviors and navigation techniques will also be described. Finally, a short description of the possible extensions of the system to connect multiple experimental setup through the Internet will close the chapter.

This chapter is based on the publication *Multi-robots Control and Tracking Framework for Bio-hybrid Systems with Closed-loop Interaction* (Bonnet et al. 2017a) with slight modifications to fit the present dissertation.

7.2 Overview of Existing Software Solutions for Animal-Robot Experiments

The analysis of existing approaches in the field of bio-hybrid systems (Swain et al. 2012; Shi et al. 2015; Landgraf et al. 2016) shows that generally researchers develop the necessary software to perform the tracking of the agents and the control of the robots in a closed-loop by themselves, as no such specific software is available commercially or open-source. These in-house software are usually not distributed, the only exception being the MADTraC library (Swain 2011) developed at Princeton University, Princeton, NJ, and that was used in Swain et al. (2012) to track a group of golden shiners (*Notemigonus crysoleucas*) and a model of a three-spined sticklebacks (*Gasterosteus aculeatus L.*) steered by a MiaBot Pro wheeled robot. MADTraC

F. Bonnet and F. Mondada, *Shoaling with Fish: Using Miniature Robotic Agents to Close the Interaction Loop with Groups of Zebrafish Danio rerio*, Springer Tracts in Advanced Robotics 131, https://doi.org/10.1007/978-3-030-16781-3_7

Fig. 7.1 The logo of the Control And Tracking Software (CATS) developed for automatic experimentation involving mixed groups composed of fish and robots

provides a desired functionality, but, unfortunately, its support is discontinued since long time with the last changes made in 2011. An option would be to use ROS (Robotic Operating System) (Quigley et al. 2009) that has a distributed and modular design, and implements its own navigation stack and bindings with the OpenCV computer vision library (Bradski 2000). However, the current version of ROS does not offer a support for multi-robot systems (ROS 2015), and it is still rather robotic research oriented, and thus potentially demanding a significant learning effort. In our case, we were willing to build software tools easy to be used by researchers in the field of behavioral biology. Also, ROS is only fully supported on the Unix-like systems, which is rather limiting, as we target biologists who are often Windows or MacOS users. Therefore, we decided to develop new software tools and target all the research field in behavioral biology, in particular those who are building mixed societies of animal and robots.

7.3 CATS Global Description

The design of the software was defined by several requirements. In our case, the software needed to perform experiments involving mixed societies of multiple agents, fish and robots, in which robots interact with animals. Thus, a robust and real-time tracking system that is closely tightened with the robots' control systems was required. The software had to be used in different types of experiments, and thus both the tracking and control parts must be easily expandable to implement desired behaviors or other functionalities. Also, the tracking part might be used separately when experimentations with only animal are run. Hence, the design must be modular to easily extract the tracking functionalities from the rest. As the software would be mainly used by biologists, the user interface must be clear, especially for the control part with a high level of abstraction. Last but not least, the software must be multi-platform.

7.3.1 Software Versions

In the first years of the project, we developed a prototype version of the software written in Python language (Bonnet et al. 2014) (Fig. 7.2a). The choice of having the

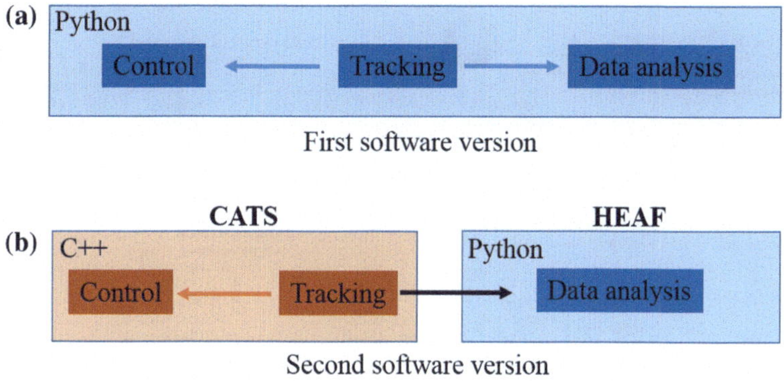

Fig. 7.2 The two versions of the software tools developed during the project to perform experiments with mixed groups of fish and robots. **a** In the first version (Bonnet et al. 2014), the tracking, control and data analysis were merged into a unique software written in Python language. **b** In order to increase the efficiency of the tracking and control parts, we separated the control and tracking from the data analysis and created two modules: Control And Tracking Software (CATS) and High throughput Ethomics Analysis Framework (HEAF)

software written in Python was twofold: first, the biologists working in the project ASSISIbf were mostly used to Python coding, thus, it was easier for the early software developments that was done in common with biologists. Also, Python offers many libraries for data analysis, tools that could be integrated into the software for automatic analysis of the data retrieved during the experiments. This software proved to work well in the first phase of the project and could manage the first fish-robot interaction experiments that are shown in Bonnet et al. (2014) and that will be described in Chap. 9.

With the complexity of the algorithms developed to perform the tracking, and the future integration of evolutionary algorithms, it was then decided to separate the control and tracking from the data analysis. The control and tracking software (CATS) would be written in C++, using efficient libraries such as Boost (Boo 2016) to reduce the computational power required, and the data analysis, that was named High throughput Ethomics Analysis Framework (HEAF) would be written in Python and mostly developed by the biologists in the University of Paris Diderot (Fig. 7.2).

7.3.2 CATS Architecture Overview

The overview of the second and most recent version of CATS software is presented on Fig. 7.3. It consists of two main components: *Tracking*, which tracks the agents (robots and fish), and *Robot control*, which generates commands to control the robots (FishBots and RiBots). The tracking and control loops are decoupled. The tracking runs at the frequency given by the Basler camera (15 Hz), and the frequency of the control loop can be configured, as it will depend on the desired behavior of the robots. CATS is implemented in C++ with extensive use of the Qt framework (Qt2 2016).

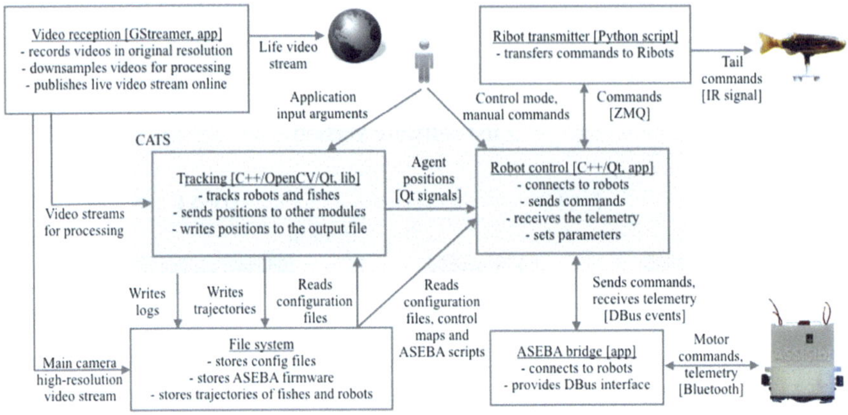

Fig. 7.3 Overview of the software architecture, used to save videos of the experiments as well as the data extracted, to track in real-time the positions of the fish and the robots, and to control the behavior of the robots. The video stream from the main camera fixed above the setup is compressed and saved on disk in high resolution (2040 × 2040 pixels). The video stream is also converted to a lower resolution (500 × 500 pixels) and published on the Internet (streamyfish.com). The tracking of the fish and lures is performed in real-time on the low-resolution video stream. The second video stream (640 × 480 pixels) from the camera fixed under the setup is used to track the FishBots. The robot control makes use of the tracked positions of the robots and fish to control the FishBots' motion as well as RiBots' body movements. The low-level control of the FishBot mobile robots is achieved by using the Aseba framework, as shown in Chap. 4

7.4 GUI

CATS provides a Graphical User Interface (GUI). It allows the experimenter to assess the progress of an experiment, visualize the tracked positions of the agents and control the robots. The different behaviors of the robots and the navigation parameters can be adapted by the user during the experiment through the GUI. The user can also visualize the robot states, for instance in case of issues such as a loss of power or if the robots are avoiding obstacles.

7.5 Video Capture and Streaming

The library Aravis (Ara 2016) is used to access the frames grabbed by the main Basler camera. All video stream operations are handled using the GStreamer library (GSt 2016). The parameters of the GStreamer media components were tuned in order to achieve very low latency. The video stream from the camera is split into two different streams: one in high-resolution (2040 × 2040 pixels, grayscale) that is saved on a disk for further analysis, the other in a lower-resolution (500 × 500 pixels, grayscale) for the image processing in CATS. The low resolution video stream is also published on the Internet in the web page streamyfish.com, so that people can observe the experiments performed online (Fig. 7.3).

The tracking part uses only the low-resolution video stream to track the positions of the agents. The use of low-resolution frames instead of high-resolution ones allows the tracking process to be less computationally expensive. We tuned the parameters of the GStreamer media components to have a very low latency. Inside CATS, the video frames transfer between different modules is done via single-producer, single-consumer lock-free queues developed by Cam (2016).

7.6 Tracking of the Fish

An overview of the tracking sub-system is presented in Fig. 7.4. The tracking is using as inputs the frames grabbed from the two cameras, the Basler camera for the tracking of the agents from the top, and the Fisheye bottom camera, for the tracking and control in closed-loop of the FishBots (Fig. 6.2).

7.6.1 *Online Tracking of the Fish*

The tracking of the agents is performed on the low-resolution (500×500 pixels) video stream. All operations are processed using the OpenCV library (Bradski 2000).

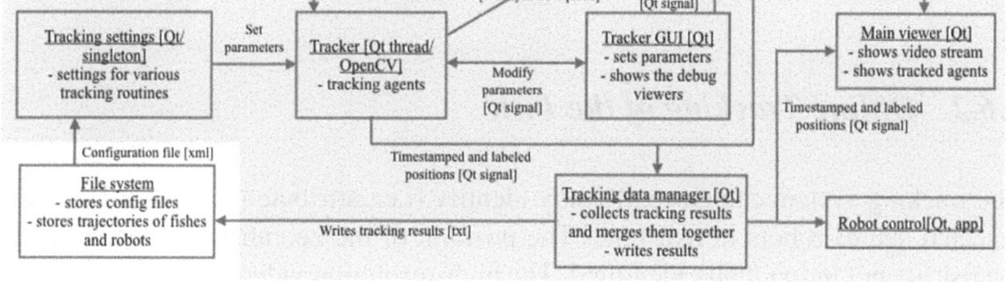

Fig. 7.4 Overview of the tracking sub-system of CATS. The stream coming from the main camera is used to track all the agents inside the arena, while the bottom camera's stream is processed to detect the robots' positions. The tracking results are later merged together to separate the fish from the lures. The resulted positions are stored in the file system for further analysis, but are also sent to the robot control sub-system. Several tracking methods are available and can be selected in the configuration file. The software provides a GUI that displays input video streams with the tracking information

First, we apply a background subtraction preprocessing step, on each frame, by using the Gaussian Mixture based Background Foreground Segmentation method of KaewTraKulPong and Bowden (2002). The position of the agents is detected by using a corner detection method (Shi and Tomasi 1994) on the resulting foreground frame, as the heads of the zebrafish and the different lures have a very sharp corner. By assigning manually the agent position at the beginning of the experiment using the GUI, this tracking method was also used to track the identified robots position. Afterwards, the tracking system updates the estimated positions of the robots by selecting the closest blob to the previous positions of the robots.

The tracking and control application was also used to analyze the locomotion of the agents. Using the positions (p_x,p_y) of the agent moving inside the aquarium, that is retrieved by the tracking, we could compute the linear speed (v_x,v_y) using

$$v_{x_t} = f \cdot \left(\frac{p_{x_{t-1}} - p_{x_{t+1}}}{2}\right), v_{y_t} = f \cdot \left(\frac{p_{y_{t-1}} - p_{y_{t+1}}}{2}\right) \tag{7.1}$$

where f is the frame rate of the image grabber, which was set to 15 frames per seconds. We computed the speed using the position difference between three frames instead of two frames in order to average the noise given by the tracking. We could obtain the linear speed of the agent v_{lin} using

$$v_{lin} = \sqrt{v_{x_t}^2 + v_{y_t}^2} \tag{7.2}$$

The angular speed ω_t was computed using the orientation difference α_t between two points:

$$\alpha_t = \text{atan}\left(\frac{p_{y_{t-1}} - p_{y_{t+1}}}{p_{x_{t-1}} - p_{x_{t+1}}}\right). \tag{7.3}$$

The angular speed is then given by

$$\omega_t = f \cdot (\alpha_t - \alpha_{t-1}). \tag{7.4}$$

7.6.2 *Offline Tracking of the Fish*

The tracking system can currently only identify (i.e., attribute the correct ID to the detected agent) robots in real time. The position of the zebrafish are detected, but the fish are not individually identified. The high-resolution videos obtained using the Basler camera are analysed off-line by the idTracker software (Pérez-Escudero et al. 2014) to identify the zebrafish (Fig. 7.5, left). This process is time-consuming and computationally intensive. For instance, when using a 32-cores computer, idTracker takes eight hours to track and identify five zebrafish in a 30 minutes high-resolution video. However, idTracker is relatively reliable: no false positive, no propagation of

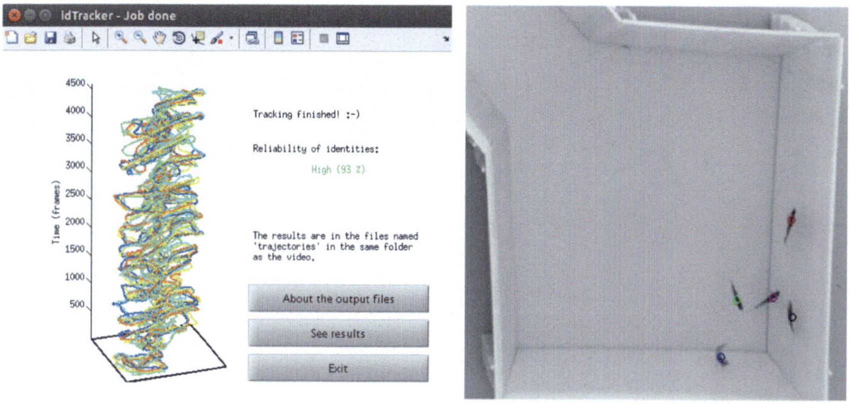

Fig. 7.5 Result obtained using idTracker (Pérez-Escudero et al. 2014) on the high-resolution video for the identification of the zebrafish. Left: The output of idTracker, with the identified zebrafish positions over time and the percentage of tracking success over the whole experiment. Right: Screenshot of the video of an experiment in which we added the position of the individual fish tracked with color circles added around the correctly identified fish. This software guarantees the identification of the fish 95% of the time, but is too heavy to run in real-time during the experiment

identification errors, and fish are identified correctly in 95% of time-steps on average. In Fig. 7.5, right, we show an example of a frame grabbed by the Basler camera, on which we added colored circles around the correctly identified zebrafish, obtained using the idTracker software.

7.7 Tracking of the Robots

Several approaches were used to detect the position and identify the FishBots. During a first period, as we were mainly making tests using one robot that was moving among fish, we used the same tracking that was implemented to detect the fish as described in Sect. 7.6.1. We could thus detect the lure the same way as we detected fish, and use the small difference of visual appearance between the lure and the fish, i.e. the size (Fig. 7.6, left), to identify the robot. This method was acceptable while we were working with only one robotic agent, but it brought some issues when working with multiple robots. First, in case a lure was decoupled from a FishBot below, it was not possible to recover the experiments, as the software had no information on the current location of the FishBots. Also, when using several agents, the loss of the identification of the robots appeared a lot due to the overlapping of the zebrafish and lures on the frame grabbed by the Basler camera (Fig. 7.6, left). Therefore, a better tracking strategies for the robot was needed to conduct mixed society experiments.

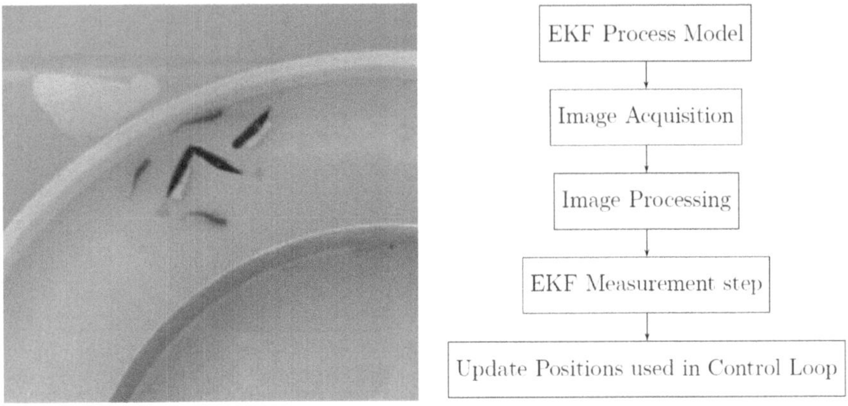

Fig. 7.6 Left: Typical issue that can occur during multi-robot experiments. Two or more lures are next to each other, which will generate a tracking issue if using a blob or edge detector. Hence, from that moment, the correct identification of the robots is not guaranteed. Right: Diagram flow of the tracking thread in CATS with the EKF algorithm added. By adding filters of error and the robots motion model into the tracking loop, we could solve some of the tracking errors

7.7.1 Kalman Filtering Approach for Improving the Tracking of Agents

An Extended Kalman Filter (EKF) was implemented to reduce the risk of losing the identification of a robot. The Kalman filter is used in many different fields and is known to be very robust and efficient (Kalman 1960). Since both fish and robots are continuously tracked and in sight, there is little risk of failure. However, the main difficulties of tracking is the overlapping of agents as shown in Fig. 7.6, left.

The EKF implemented on the tracking can be seen in the form of a flowchart in Fig. 7.6, right. The process model estimates the future pose, i.e., the position and orientation, of the robots using the motion command that is sent to the robots, then the image tracking defined in Sect. 7.6.1 is used in the measurement step of the filter.

For the process model, we used the kinematic model of the FishBot to predict its next state. The FishBots are moving using a differential drive configuration of the wheels located in the center of the robot. We begin by defining the equations of motion of a differential drive mobile robot in function of the tangential speeds of each wheel

$$v_{lin} = \frac{v_{left} + v_{right}}{2} \tag{7.5}$$

$$\omega = \frac{v_{right} - v_{left}}{l} \tag{7.6}$$

where v_{lin} is the linear speed, ω the rotational speed, v_{left} and v_{right} the tangential speed of the left and right wheel, respectively, and l the distance between the two wheels which is 2 cm for the case of the Fishbot.

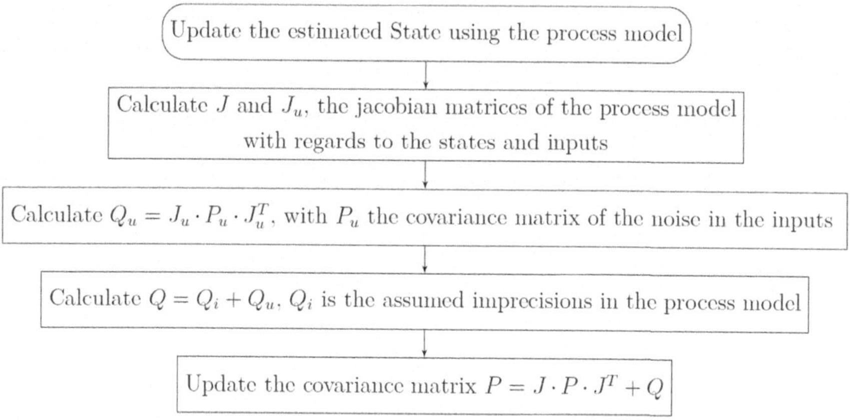

Fig. 7.7 Flow chart of the EKF process model

The flow chart of the process model of the EKF for one iteration can be seen in Fig. 7.7. The process model used in the Kalman Filter can also be written as

$$x_{t+dt} = x_t + dt \cdot v_{lin} \cdot cos(\phi_t) \tag{7.7}$$

$$y_{t+dt} = y_t + dt \cdot v_{lin} \cdot sin(\phi_t) \tag{7.8}$$

$$\phi_{t+dt} = \phi_t + dt \cdot \omega \tag{7.9}$$

The Jacobian matrix of the process model is given by

$$J = \begin{pmatrix} 1 & 0 & dt \cdot v_{lin} \cdot sin(\phi_t) \\ 0 & 1 & -dt \cdot v_{lin} \cdot cos(\phi_t) \\ 0 & 0 & 1 \end{pmatrix}$$

For this implementation, the process model uses the speed command that is sent to the FishBot and not the inputs from the encoders as it is usually done for such system. The reason for this is that the FishBots are capable of closely follow the given commands as it can be observed in Fig. 4.15, as long as the speed command stays below 30 cms^{-1}. Also, the Aseba network that is used to exchange the events containing the data between the FishBots and CATS would probably saturate if the robots sensing information are regularly sent to CATS, especially if a high number of robots is used. The fact that we use the commands sent to the robots as input to the filter reduces its accuracy. In order to compensate for this, we assume a high noise in the inputs, which is done in the matrix J_u. However, in reality, this noise is non-Gaussian as it is in fact a time delay. This means that we cannot easily compensate the error with a Kalman Filter.

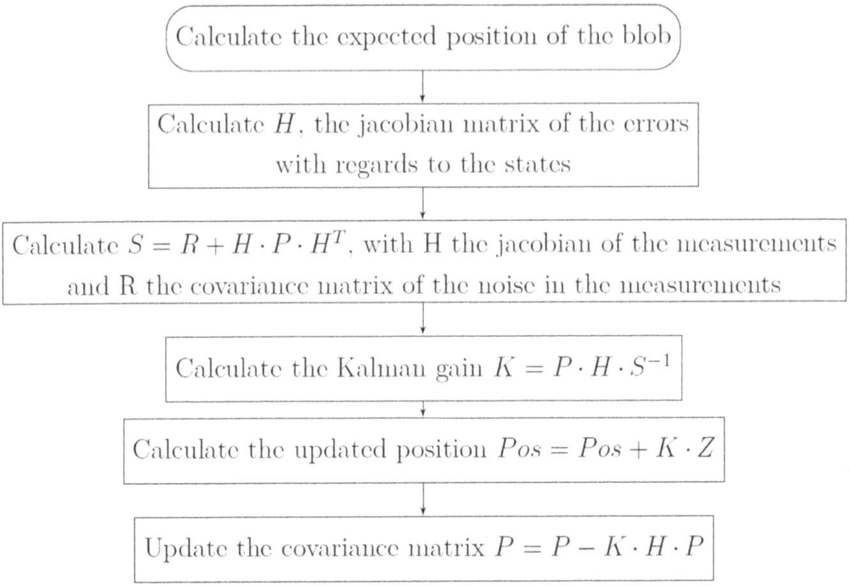

Fig. 7.8 Flow chart of the measurement process of the EKF

The matrix J_u is given by:

$$J_u = \begin{pmatrix} 0.5 \cdot dt \cdot cos(\phi_t) & 0.5 \cdot dt \cdot cos(\phi_t) \\ 0.5 \cdot dt \cdot sin(\phi_t) & 0.5 \cdot dt \cdot sin(\phi_t) \\ -dt \cdot \frac{r}{l} & dt \cdot \frac{r}{l} \end{pmatrix}$$

where r is the wheel radius which is 5 mm for the case of the FishBot.

The values for the standard deviations used could be edited in the configuration file in order to reduce the need for recompiling the software.

The measurements used in the measurement process are the position p_x and p_y of the lure obtained using the same method described in Sect. 7.6.1. The measurement process is described in Fig 7.8.

Using the Kalman filter for the tracking, we first noticed that the tracking accuracy was not improved at low speed, but increased at high speed. This is due to the fact that at low speed, the method to identify the head of the lure is very efficient, but its efficiency decreases at high speed. As the Kalman Filter is including the motion of the robot into the estimation of the trajectory, it results in a slightly improvement of the tracking at high speed up to 30 cms^{-1}. Above 30 cms^{-1}, the EKF is offering poor results, due to the fact that, at these ranges of speed, the robot is not behaving as encountered and do not follow exactly the desired speed.

However, the types of problems that are shown in Fig. 7.6 where not fully resolved, due to the fact that the measurements from the edge detector methods were always taken into account, even when two robots were close to each other. We thus tried to implement some supplementary conditions into the filter. For instance, we made the tracking ignore the measurement steps when two or more robots were too closed to each other. Unfortunately, the process model could not guarantee that the robots would not lose their IDs, mainly because when entering into collision, the process model was not able to accurately estimate the poses of the robots.

Due to the fact that the Kalman filter could not guarantee a 100% robot identification performance, we decided to implement another solution for the tracking that is not only based on the image retrieved by the top high-resolution camera.

7.7.2 FishBot Tracking from Below

It thus occurred that the best way to ensure the identification of the robots during the experiments was to have a second solution of tracking coupled with the one using the camera on the top, as it is done in Landgraf et al. (2013). However, compared with the solution of Landgraf et al. (2013) in which they use a transparent support for the mobile robot, the support on which the FishBots are moving needs also to provide the powering of the robot. Thus, a solution to have a support able to power the system and for the robot to be seen from below had to be found.

The first adopted solution was to modify the copper conductive plate below the FishBots by a see-through conductive material. Indium Tin Oxide (ITO) coating on a glass was investigated, but tests showed that the coating had a too high resistivity. We tested a second solution involving a thin PCB with a copper grid. A first prototype of this solution was implemented using thin PCB substrate sheets of 0.1 mm made of FR4. The PCB was covered by a thin copper grid of 35 μm. As the FR4 is translucent with this thickness, LEDs mounted on the bottom of the FishBots could be seen through. We could perform some experiments using this setup to validate the method. However the solution had several disadvantages: First it was not possible to print a PCB sheet large enough to cover the entire setup, thus we had to assemble several sheets and bridge them to ensure conductivity, but then the flatness was not guaranteed. Also, over time, the copper grid would be damaged due to friction and oxidized.

The final solution adopted involves the use of a 1 mm thick stainless steel plate with small holes that was already introduced in Chap. 6. The dimensions of the holes are 1 mm diameter, distant of 2 mm and disposed in a triangular shape. The size of the holes is small enough so that it does not perturb the motion of the FishBots, and the lights coming from the LEDs can diffuse through the holes (Fig. 7.9, left). The stainless steel plate is placed on a Plexiglas support to not bend. Frames are grabbed from below by a fisheye camera with a 180° field of view. Each FishBot is equipped with six LEDs of the same color, three located in the front and three in the back (Fig. 7.9, right). A blob detector is performed on the HSV color space images to localize the position of the LEDs, and the position of the robot is estimated in the middle of two blobs of the same color. This solution is very robust, as it is not possible to lose a robot, as no occlusions appear, and each robot can be identified by its corresponding LED color.

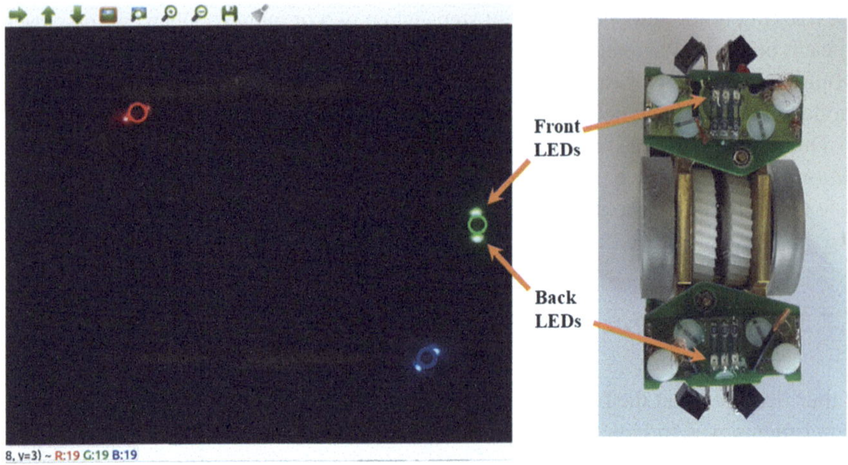

Fig. 7.9 Tracking of the FishBots from below. Left: View from the fisheye camera of three FishBots from below, each equipped with six LEDs with the same color for identification. A circle is added around the estimated center of each FishBot. Right: The FishBots seen from below, with the two PCBs equipped with 3 LEDs each

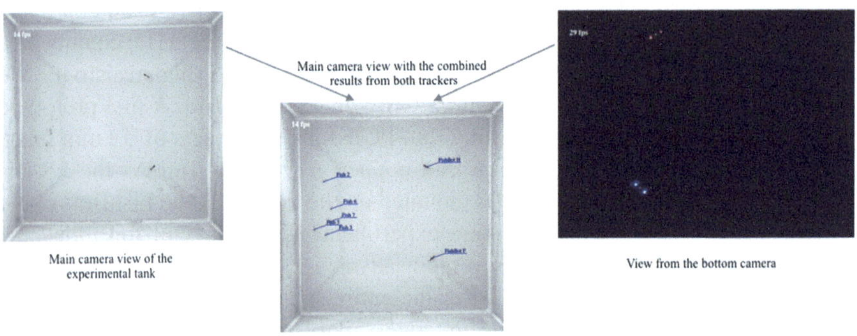

Fig. 7.10 The merge of the tracking results from (left) the camera above the setup that tracks all the agents, lures and zebrafish, and (right) the camera below the setup that tracks only the FishBots. The tracking results from both cameras are converted from pixels to millimetres thanks to the camera calibration routine and then merged together to detect the robots among all the agents

We implemented this solution inside CATS to merge the two tracking methods (Fig. 7.10). First, we calibrated the cameras by removing the distortion using a chessboard and OpenCV methods to retrieve the distortion matrices. Then, we placed in the position of the FishBots a specifically designed grid made of LEDs at the bottom and magnets at the top. We measured the positions of each nodes of this grid on the two various framed grabbed by the Basler camera and the fisheye camera. These two matrices are used to merge the two frames to obtain the same pose of the FishBots in the global frame.

7.8 Control of the FishBots

7.8.1 Overview of the Control Architecture

The modular event-based architecture for the mobile robots library (Aseba) has been used to individually control the FishBots in real-time and reprogram them during an experiment without flashing the firmware of the microcontrollers (Sect. 4.6.3). The control of the FishBots' motion is done through events that are sent from CATS and that contain the parameters for the locomotion. Some behaviors are implemented onboard each FishBot or at the level of CATS. Thanks to the event-based protocol, the FishBots are able to emit events in case of obstacle presence or powering issues, and the control application can then modify the behavior of the robots to overcome these types of situations.

The robots are thus connected to the same Aseba network using Medulla (Sec. 4.6.3). In order to interface with Medulla, a DBus interface was implemented in CATS. The interface allow to get all the functionalities of Aseba in CATS, such as loading Aseba scripts in the virtual machine onboard each FishBot, getting and sending events, modifying Aseba variables etc. This software module can be found as a contribution to the Aseba community in QtD (2016).

Table 7.1 shows a visualization of all the behaviors that were implemented in the firmware of the FishBots and in CATS in order to conduct the required experiments of ASSISIbf (Fig. 7.11).

7.8.2 Obstacle Avoidance Mechanisms

As we are using arenas that can be composed of corridors or small rooms (see Chap.6), automated obstacle avoidance mechanisms were required to avoid the collisions of the FishBot with the boarders of the arena or with other FishBots, which could generate

Table 7.1 List of the FishBots' behaviors. Target generator is the mechanism that generates a goal for the FishBots. The locomotion patterns is the locomotion that is used by the FishBot when moving toward the target. Obstacle avoidance mechanisms are used during the displacements of the FishBots. Path planning algorithms were also implemented to find the optimal trajectory, taking into account the shape of the arena, the positions of all the robots, and the position of the targets

Target generator	Locomotion pattern	Obstacle avoidance	Path planning
Joystick manual control	Constant linear speed	Braitenberg	Dijkstra
Pre-programmed	Fish-like patterns	Turn and go	Potential field
Fish shoal centroid	Straight ahead	Potential field	
Vision-based model			

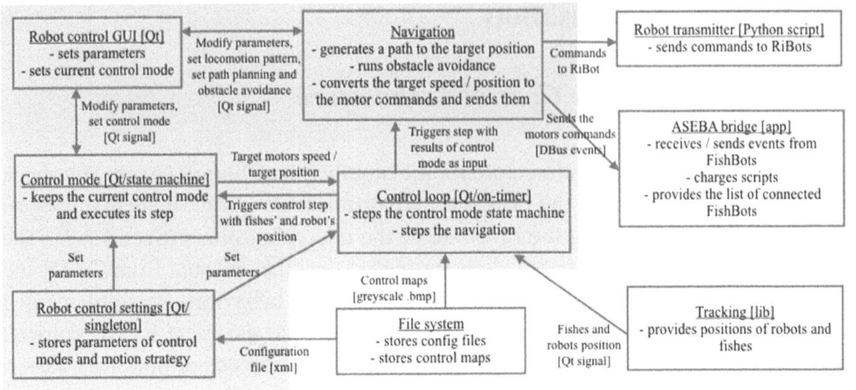

Fig. 7.11 Overview of the robot control sub-system of CATS. Several kinds of control modes (behaviors) and locomotion patterns are available, and can be selected by the user in the user interface during an ongoing experiment. For the FishBots, the desired speed of each wheel is sent through a serial connection via the Aseba network, in which all the FishBots are connected and can receive or emit events. For the RiBots, the communication is only one way, with the IR RC5 signal that is broadcast to control the stepper motors of the RiBots that drive their tail

a failure of the experiment, with a FishBot blocked or a decoupling between a lure and a FishBot. We chose to implement these algorithms on two software layers: some at the level of the FishBots, which were already described in Chap. 4 that are based on the proximity sensors measurements retrieved by the FishBots and some at the level of CATS that are taking into account the global position of all the agents and the shape of the arena.

Local Obstacle Avoidance

If an obstacle is detected by a FishBot, it will first send an event to the Aseba network to inform that it has detected an obstacle. This information will be retrieved in CATS to inform the user that this FishBot has detected an obstacle and also, depending on the current behavior of the robot, CATS will automatically stop sending control commands to this robot while it is avoiding the obstacle by itself. The robot will then avoid the obstacle autonomously. As explained in Sect. 4.7, two types of algorithms were implemented: One based on the Braitenberg vehicle principle (Braitenberg 1986) and one named Turn and Avoid which allows the robots to escape any situation that involves obstacles.

High-Level Obstacle Avoidance

The local obstacle avoidance methods proved to resolve most of the problematic situations for experiments involving up to three FishBots. However, some situations occurred in which three FishBots were blocked for 10–20 seconds in a small corridor or in the circular arena for instance (Fig. 7.6, left), trying to avoid each other and slowly finding a solution to escape this situation autonomously. This is of course not acceptable for experiments in which we want to show that the robots are mimicking

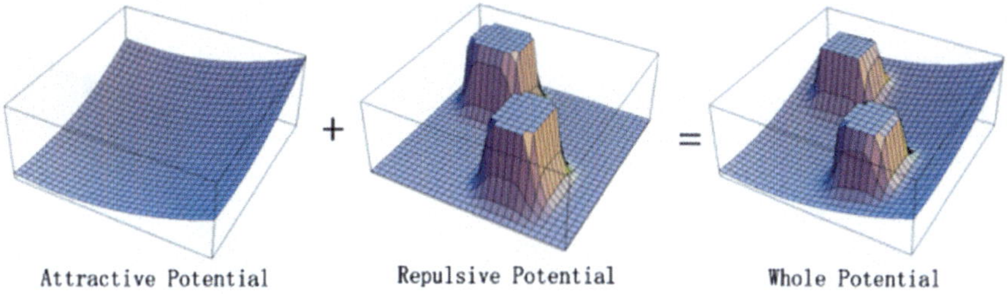

Fig. 7.12 Overview of the potential field algorithm principle. On the left, an attractive potential, for which for example the robot will move towards the left to reach a target located on the corner of the squared arena. In the middle, a repulsive potential field with two obstacles that the robot has to avoid. On the right, the combination of the attractive and repulsive potential fields which conduct to a path free of collisions for a mobile robot (Wang 2017)

correctly the zebrafish behavior. Also, knowing exactly the environment and the position of the agents thanks to the tracking could allow to design more sophisticated algorithms.

Hence, we implemented an obstacle avoidance mechanism at the level of CATS based on the potential field algorithm (Khatib 1986). This algorithm is based on the idea of applying virtual forces onto the robot and construct an artificial potential field. The obstacles repel the robots and the targets attract the robots which will therefore be able to navigate toward the goals, located at the global minimum of the potential field (Fig. 7.12). The advantage of these types of path planning algorithms is that they are easy to implement and can also incorporate sensory inputs.

As the potential field algorithm involves many parameters, such as the shape of the attractive and repulsive potential fields, which are not trivial to tune, preliminary parameter tuning was conducted using Matlab (Fig. 7.13). After that, the algorithm was implemented in CATS, and tests involving two robots could demonstrate that this algorithm can be used for instance in the two rooms setup (Fig. 6.5b) to increase the collision avoidance efficiency between two robots, as the algorithm takes into account the arena structure and the position of the robots.

7.8.3 Target Generators

In the framework of the ASSISIbf project, we are studying the effect of various stimuli that can be generated by the FishBots on the zebrafish. The locomotion patterns generated by the FishBot, as well as the types of trajectories that are followed by the devices, are typical stimuli that are of great interest to study the social behavior of fish. We have implemented in CATS different types of target generators and locomotion patterns that the user can either select on a configuration file, or adapt manually during an ongoing experiment. The target is the goal that the FishBot has to reach, and the locomotion pattern is how the robot will move to the target.

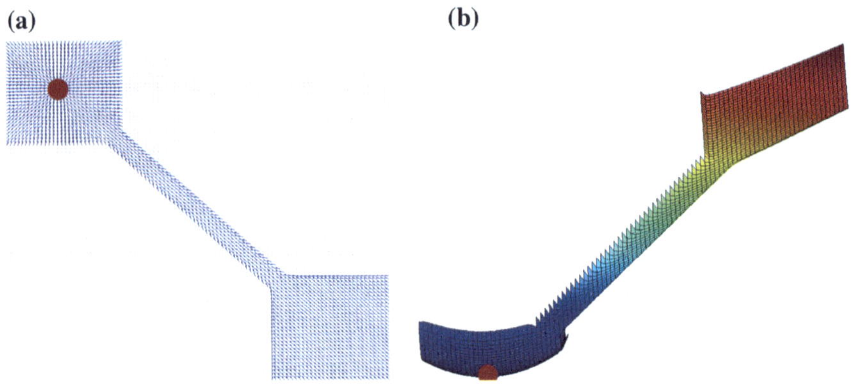

Fig. 7.13 Simulation of the potential field algorithm applied on the two rooms setup using Matlab. The forces and potential field due to the repulsion of the arena and the attraction of the target are represented at each position of the aquarium, with visible gradient descent to the red target located at the local minimum. **a** 2D Plot showing the direction of the forces. **b** 3D representation of the gradient

Manual Mode

For debug purposes and in order to initialize the system, we have implemented a manual mode to control the Fishbots. Each FishBot can be move manually using a joystick. The experimenter can also manually place a target using the mouse.

Preprogrammed Trajectories

The user can define in a configuration file the trajectory of each FishBot. For instance, this mode was used in order to generate a specific trajectory in the open arena setup that was used to perform the preliminary experiments involving the FishBot with zebrafish that are described in Chap. 9. The targets are simply the nodes of the trajectory and are updated either depending on the proximity of the FishBot or at each time step with a frequency that can be configured.

Fish Shoal Centroid

In this mode, the target is generated in the centroid of the fish shoal. This behavior was tested to observe how the zebrafish would react to robots moving continuously in the center of their shoal as it is done in Swain et al. (2012), Landgraf et al. (2013, 2016). However, this behavior is not mimicking the fish behavior, as the fish do not always swim in the direction of the shoal centroid. Thus we mostly used the Vision-based model that will be further described.

Vision-Based Model

The zebrafish displacement model described in the next chapter (Chap. 8) is used to generate a target based on the current pose of simulated or real agents in the arena, as well as the walls of the arena. It allows for instance a FishBot to mimic the fish behavior such as shoaling, by being attracted by the position of other agents or the walls, which is a behavior observed on our zebrafish (Collignon et al. 2016). A more detailed description of the model is done in Chap. 8.

7.8.4 Locomotion Patterns

Straight Ahead

For the straight ahead mode, the FishBot will move on a straight line and a parameterized speed is continuously sent to it so that it follows a straight forward trajectory. If coupled with one of the local avoidance method, it will generate a random walk of the robot inside an arena.

Constant Linear Speed

The FishBots follow the target at a parameterized linear constant speed. This controller was implemented in order to mimic the zebrafish motion inside corridors, such as in the circular corridor arena, or the two rooms arena. The control of the rotational speed is done using a PID based on the difference between the current pose of the FishBots and the target position.

The angle difference between the position of the robot and the position of the target that it has to reach is defined by:

$$\beta(t) = \text{atan}\left(\frac{dy(t)}{dx(t)}\right) \tag{7.10}$$

where $dx(t)$ and $dy(t)$ are the difference between the current robot position and the position of the target along the x and y axis, respectively.

The difference between the current orientation of the robot $\rho(t)$ and the angle between the position of the target and the robot's position given by $\beta(t)$ is the angle $\alpha(t)$ that the FishBot has to turn to reach the target

$$\alpha(t) = \rho(t) - \beta(t) \tag{7.11}$$

The controlled rotational speed ω_c is computed using a PID controller based on the computed angle $\alpha(t)$

$$\omega_c = K_p\alpha(t) + K_i\int_0^t \alpha(\tau)d\tau + K_d\frac{d\alpha(t)}{dt} \tag{7.12}$$

where K_p, K_i and K_d are the proportional, integral and derivative gain, respectively. The gains of the PID can also be parameterized, as they will depend on the linear speed of the robots, or the type of arena that we are testing.

Using Eq. 7.6, and in function of the ω_c computed using the PID formula in Eq. 7.12, we obtained the following controlled tangential speeds of the left and right wheel ($V_{left,c}$ and $V_{right,c}$, respectively)

$$V_{left,c}(t) = V_{l,p} + \frac{\omega_c(t) \cdot l}{2}, V_{right,c}(t) = V_{l,p} - \frac{\omega_c(t) \cdot l}{2} \tag{7.13}$$

where $V_{l,p}$ is the parameterized linear speed and l is the distance between the two wheels of the FishBot.

Fish-Like Patterns

The Fish-like pattern is a finite state machine implemented at the level of the FishBot that follows a sequence of three states: orientation, acceleration and relaxation, which mimics the zebrafish locomotion underwater, while following a given target. This controller is described in detail in Chap. 8.

Adaptive Fish-Like Patterns

The adaptive fish-like patterns follows the same controller as the fish-like patterns, but the latter used preprogrammed parameters to generate the speed profile of the locomotion patterns, as, for the case of the adaptive fish-like patterns, the parameters are adapted in CATS depending on the current pose of the FishBot. Indeed, we measured that the locomotion patterns of zebrafish could depend on their location in the arena, thus, the robots had also to adapt to these changes of locomotion patterns. The measurements performed to compare the FishBot and zebrafish locomotion, as well as the implementation of the adaptive fish-like patterns are also described in detail in Chap. 8.

7.8.5 *Path Planning*

Two algorithms were implemented in order to increase the efficiency of the navigation, by finding the optimal trajectory that the robots has to follow to reach the targets. One of them is the potential field algorithm that was already described in Sect. 7.8.2. The other one is the Dijkstra's shortest path algorithm.

Dijkstra's Shortest Path Algorithm

The Dijkstra's algorithm finds the shortest path between nodes in a graph (Dijkstra 1959). The graphs are made of two elements: nodes and vertices. The nodes are the accessible points in the graph and the edges represent the paths connecting each pair of nodes. This algorithm can thus be used to find the shortest path between the current location of the robots and the targets, by taking into account the shape of the arena. It is a very suitable algorithm for complex arena, such as the two rooms arena (Fig. 6.5b) or more complex arenas that were considered in the ASSISIbf project. As the algorithm had already an implementation in the boost library (Boo 2016), we also integrated this algorithm in CATS. This algorithm proved to increase the efficiency for a FishBot to navigate more smoothly in the two-rooms setup during the transition between the two rooms, as it will considered for instance the shape of the entrance in the corridor.

7.9 Control of the RiBots Lures

For the RiBots, a Raspberry PI, on which Linux Infrared Remote Control (LIRC) library is running (lir 2017), is connected through ethernet with the main computer (Fig. 6.2), and RC5 signals are generated on an output pin connected to an IR emitter. The IR signal is broadcasted over the whole aquarium and received by all the RiBots (Fig. 6.2). An ID is included inside the control command in order to individually control the RiBots.

7.10 Long-Distance Infrastructure

As the experiments involving mixed societies of fish and robots could take place in different sites during the project, Lausanne, Paris and also other places in Europe, it was also required for the software infrastructure to be able to work remotely with experimental setups located in another places than the experimenter.

The Aseba interface to control the robots has demonstrated to allow the control of robots that are located at a long-distance through the Internet (Mondada et al. 2016). We showed in Sect. 7.5 that our infrastructure allows the streaming of video flux, and we can retrieve these stream in CATS. Therefore, we are able to launch CATS from long distance with a setup not located at the same place CATS is running. We are also able to connect different setup that are located in different places as shown in Fig. 7.14.

7.11 Conclusion and Contribution to the State of the Art

We introduced a novel software framework to perform automatic long-duration experiments with a mixed society composed of several fish and several robots. Moreover, CATS is highly modular, flexible and efficient. It is able to track zebrafish and robots in real-time, and to control several robots to exhibit reactive and complex fish-like behaviors. Our software is also designed to be generic, and could be potentially used to perform behavioral experiments in other setups, involving other animal species and other robotic designs.

The framework presented in this chapter has already been implemented into two universities for research involving mixed societies of animals and robots (University

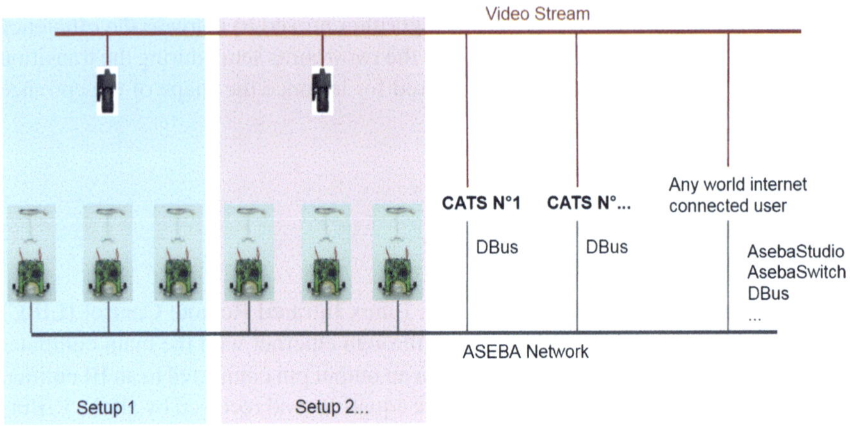

Fig. 7.14 Schematic of the long distance software infrastructure support of CATS. For the control side, all the FishBots are connected to the same Aseba network, thus they can be controlled from any places that can access this network through the Internet. Also, the Gstreamer library (GSt 2016) supports high-definition streaming over the Internet, thus CATS can be launched in another location than where the experiments is running and still be able to control the robots and retrieve biological datas

Paris Diderot in France, and Ecole Polytechnique Fédérale de Lausanne in Switzerland). It was also implemented during the ARS-ELECTRONICA 2016 festival in Linz, Austria (ars 2016) where the mixed society of zebrafish and robot was presented to the public. Finally, CATS is also open-source and is available on GitHub (CAT 2016).

7.12 People Who Contributed to This Work

The software developments was the work of mainly three contributors, Leo Cazenille, Dr. Alexey Gribovskiy and myself. Dr. Marcelo Elias de Oliveira also made an early version of the software. Dr. Stefan Witwicky contributed on the developments of the Aseba-QtDBus interface. Patrick Bobbink implemented the first prototype of tracking from below and contributed to the integration of the Extended Kalman Filter in CATS. Laila El Hamamsy worked on the implementation of the navigation algorithms, such as Dijkstra and the potential field algorithm.

Chapter 8
Biomimetic Behavior Models for Controlling a Robotic Fish

8.1 Summary

In this chapter, we present different methods developed to control the FishBot so that the lure can mimic the zebrafish movements (Fig. 8.1). In particular, we present a zebrafish-like locomotion pattern implemented at the level of the FishBot. A model capable of reproducing the trajectories made by zebrafish inside of the shoal is also presented. We conclude the chapter by showing the validation of the control algorithms in an experiment involving multiple robots and multiple fish, and showing that the trajectories as well as the locomotion patterns do mimic the zebrafish ones.

This chapter is based on the publication *Design of a Modular Robotic System that Mimics Small Fish Locomotion and Body Movements for Ethological Studies* (Bonnet et al. 2017b) with slight modifications to fit the present dissertation.

8.2 The Need of Biomimetic Controllers

As it will be further demonstrate in this dissertation (see Chap. 9), the motion of the robot is one of the key aspects to increase the acceptance of the lure and the ability to modulate the fish social behaviors. Therefore, the design of biomimetic controllers at different level of the robot control architecture was needed.

In Chap. 2, Sect. 2.5, we showed that researchers that are studying fish-robot interactions have not yet implemented complex biomimetic models to drive their robots. Thanks to the infrastructure that was developed in this study, with the miniature and dynamic mobile robot FishBot, the automated experimental setup and the software infrastructure CATS, we can achieve a closed-loop control of multiple robots following models that are mimicking the trajectories and locomotion patterns of the zebrafish.

F. Bonnet and F. Mondada, *Shoaling with Fish: Using Miniature Robotic Agents to Close the Interaction Loop with Groups of Zebrafish Danio rerio*, Springer Tracts in Advanced Robotics 131, https://doi.org/10.1007/978-3-030-16781-3_8

Fig. 8.1 Top view of a lure moving inside the two rooms arena among four zebrafish while mimicking the fish locomotion patterns

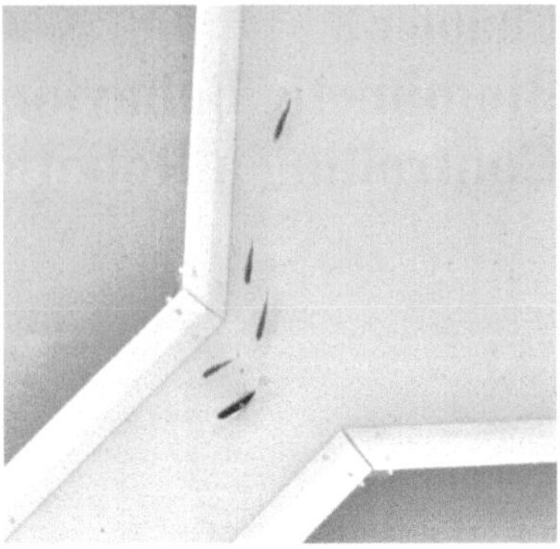

Two types of control layers were required for the robot: high-level controllers that mimic the fish trajectories, thus reproducing the shoaling behaviors inside the different types of arena that we designed (Sect. 6.3), that are using the current pose of the agents to generate target positions for the robots, and low-level controllers that mimics the fish locomotion when moving towards these targets. Also, coupled with these controllers, robust obstacle avoidance algorithms were needed to make the robots avoid collisions while mimicking the fish behavior around obstacle or other fish.

8.3 Locomotion Pattern Embodiment

8.3.1 Already Existing Controllers

Regarding the embodiment of fish-like locomotion on a robotic device, few models of fish locomotion can be found in the literature. In Mwaffo et al. (2014), fish locomotion is modelled using a jump persistent turning walker model motivated by the sudden and drastic changes in the locomotion of zebrafish in the form of large deviations in turn rate. In Zienkiewicz et al. (2015), a stochastic model is used to reproduce zebrafish locomotion in a confined environment. While these models accurately reproduce the motion of fish, the translation of their mathematical expression into concrete commands for a robot were not validated on a real system.

8.3.2 Zebrafish Locomotion Analysis

We started by extracting the locomotion of our zebrafish in order to design the robot controller. Zebrafish were transferred from their housing aquarium to the experi-

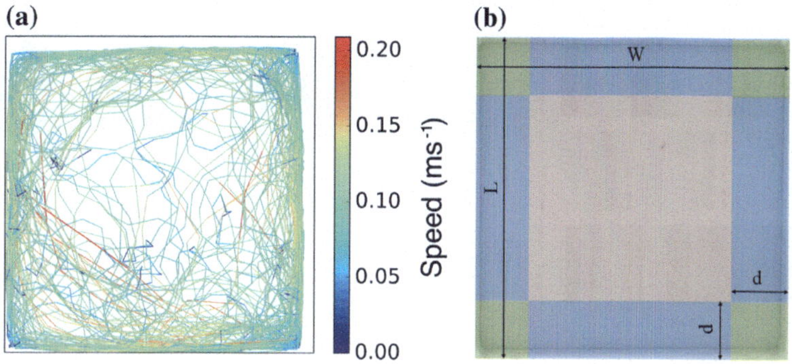

Fig. 8.2 **a** The trajectory of a zebrafish swimming in the experimental tank for one hour. The color of the trajectory indicates the linear speed of the fish at a given position ($\Delta t = 1$ s). The fish is mainly observed along the wall of the tank and decreases its speed in the corners of the tank, while higher speed values are measured in the center of the aquarium. **b** Zone delimitation in a tank of L (100 cm) and W (100 cm) for the three different types of fish behavior extraction: corners (green) that are delimited by the right angle walls of the tank and virtual lines at a distance *d* (15 cm) from the walls, edges (blue) delimited by the setup wall and the same virtual line, and the center (red). The motion patterns of the fish were classified depending on the zone in which it is swimming

mental aquarium to extract the characteristics of their trajectories over time in the open arena (Fig. 6.5a). We recorded at 15 frames per second the position of each fish swimming in the tank for one hour. Thanks to the individual tracking obtained using idTracker (Pérez-Escudero et al. 2014), we retrieved the trajectory of each fish and computed their speed and acceleration during the entire experiment. An example of a trajectory is given in Fig. 8.2a and highlights that the fish were mainly swimming along the walls of the tank, as also shown in Zienkiewicz et al. (2015). We could also identify three speed patterns of the fish according to their location in the experimental tank. Indeed, their linear speed decreases in the corners of the tank when they change their direction, while the zebrafish swim at a higher speed along the edges of the tank. Finally, we measured the highest linear speed values in the middle of the aquarium. Thus, we divided the experimental tank into various zones: corners, edges and the center (Fig. 8.2b). The distance *d* delimiting the edges and the corners was estimated at 15 cm.

As our goal is to develop a modular robotic system that mimics fish locomotion, we also performed a more detailed analysis of the motion pattern of the zebrafish. The speed of the zebrafish can be decomposed into successive cycles, starting with a sharp acceleration followed by a slow deceleration until the next tail beat (Fig. 8.3a–b, left). With regard to the direction of the fish, the changes in orientation, identified by a high angular speed, are mainly detected at the beginning of the cycle (Fig. 8.3c, left). Thus, zebrafish usually move by following a sequence of three steps:

- Step 1: **Orientation**. Strong caudal peduncle bending to reorient and start the propulsion toward the next goal.
- Step 2: **Acceleration**. High linear acceleration following the given caudal peduncle beat to reorient.

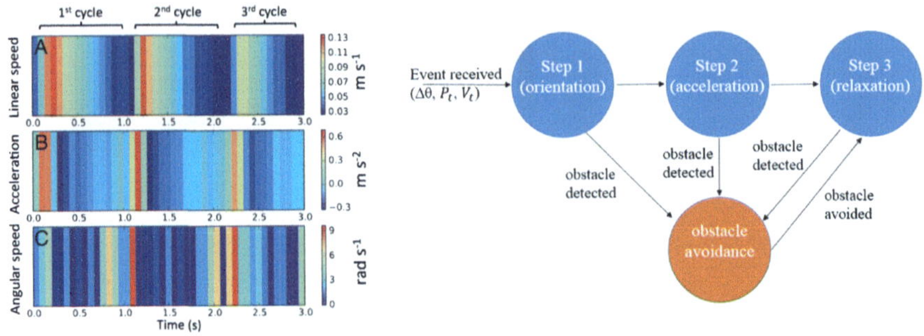

Fig. 8.3 Left: Example of the swimming movements of a zebrafish magnified for three seconds. Each colored segment represents the linear speed (**a**), acceleration (**b**) and angular speed (**c**) of the fish measured at a given position ($\Delta t = 1/15$ s). The linear speed of the fish can be decomposed into cycles that start by a sharp acceleration (hot colors) followed by a longer deceleration (cold colors). The angular speed highlights that changes of orientation occur mainly at the start of the cycles. Right: Finite-state machine implemented on the control layer of FishBots. A motion cycle is started when an event is received from the high-level control application CATS. The parameters contained on the event are used in the different steps: $\Delta\theta$ is the angle difference between the current orientation of the FishBot and the orientation needed to reach the next target, P_t is the target distance to accelerate, and V_t is the target linear forward speed. If an obstacle or another robot is detected during a cycle, a simple obstacle avoidance behavior is implemented to avoid it until a new event is received

- Step 3: **Relaxation**. Tail beating stops and the fish starts sliding into water with its linear speed decreasing.

8.3.3 Implementation on the FishBot

We implemented a finite-state machine into the FishBot firmware in order to execute the same locomotion sequence as the zebrafish (Fig. 8.3, right). An Aseba event containing the ID of the FishBot is emitted with a parameterized frequency from the control application and sent to the Aseba network on which all the FishBots are connected. When an event is received by a FishBot, it starts executing the finite-state machine with the three locomotion steps described: orientation, acceleration and relaxation. In case of an obstacle detected by the IR proximity sensors, the execution of the locomotion is stopped and the robot starts avoiding the obstacle, and then return to the relaxation state where it waits for a new event.

For Step 1, the orientation, we used the wheels position control to reorient the robot toward the target with an orientation difference of $\Delta\theta$. We used one wheel of the robot to move forward while the other wheel remains stationary. This generates a rotation of the robot with a small linear speed, as zebrafish almost never have a null linear speed even while turning.

For Step 2, the acceleration, as we have no direct control over the acceleration (Fig. 4.11), we measured the acceleration obtained using the position control in cascade with the speed and current control. By using the cascade controller, the motors are protected against overheating or too high speeds, and we can adjust the position controller gains to obtain very high acceleration. The measures are shown in Fig. 4.15, left; we can see that the increase in acceleration is linear with the target position P_t ordered up to 6 cm, where the maximum FishBot acceleration of 1.3 ms^{-2} of average was achieved. Beyond this value, the motors entered in saturation due to the limit of the system. The linear curve of this figure was used to match the acceleration of the FishBot with the acceleration of the zebrafish. We have also characterized the limits of the system in terms of speed (Fig. 4.15, right). We can observe that for speeds of up to 20 cms^{-1}, the response is relatively smooth and stable. For higher speeds, the robot takes more time to reach the desired command, and, over 30 cms^{-1}, the limit of the system is achieved. Finally, in terms of angular speed, the FishBot alone is able to achieve angular speed of up to 30 $rads^{-1}$. However, the magnetic coupling between the RiBot and the FishBot was lost over speeds of 18 $rads^{-1}$.

Finally, for Step 3, the relaxation, we used the speed controller to generate a linear forward speed V_t of the robot, which decreases over time at a rate of 1 cm every 100 ms (10 cms^{-1}), thereby reproducing the deceleration of the zebrafish.

8.3.4 Experimental Validation Using the FishBot Coupled with the RiBot

Regarding the matching of locomotion, we tuned the parameters of our fish locomotion behavior implemented on the FishBot to match the data extracted from the zebrafish locomotion patterns. The final parameters of the FishBot are shown in Table 8.1. These parameters were tuned in order to match the average linear speeds and, secondarily, the locomotion sequence curve in terms of speed and acceleration (Figs. 8.4, 8.5 and 8.6).

We recorded at 15 frames per second the position of the RiBot moving in the tank for one hour using the same experimental setup and tracking software used to analyze zebrafish movements. The FishBot was programmed to follow a trajectory similar to the zebrafish by following the walls and, from time to time, moving rapidly in the center of the tank.

Table 8.1 The parameters of the locomotion of the FishBot for the three zones of the experimental setup. F is the frequency at which the Aseba events are sent to start a motion cycle. P_t is the target position to mimic the acceleration phase, and V_t is the starting linear speed for the relaxation phase

Zone	F [Hz]	P_t (Step 2) [cm]	V_t (Step 3) [cms^{-1}]
Corners	5	2	7
Edges	8	2.7	12
Center	7	4.2	17

8.3.5 Results and Discussion

The distribution of the linear speed measured in the three zones of the tank (corners, edges, center) are shown in Fig. 8.4 for the lure and the zebrafish. As suggested by preliminary observations, the fish swim with a higher speed near the edges (0.054 $\pm$ 0.032 ms^{-1}) than in the corners (0.066 $\pm$ 0.034 ms^{-1}) of the tank, and even faster in the center of the aquarium (0.073 $\pm$ 0.031 ms^{-1}). By adjusting the parameters of the controller, we were able to reproduce similar distributions between the lure and the zebrafish. However, the fitting quality of the speed distribution is lower in the case of the lure for motions along the edges and in the center, which is not the case for the zebrafish. This can be explained by the fact that for high accelerations that are applied in order to achieve high average speeds, the lure is less regular than the zebrafish due to the friction of the lure module, the slipping of the FishBot wheels, and the inertia of the system.

In addition, with the developed controller, the RiBot was able to reproduce the sinusoidal shape of the speed and acceleration of the fish (Fig. 8.5). This is made possible by fitting the movement patterns of the fish with the movement patterns of the robots. By identifying successive local minima in the linear speed of the fish, we computed the average speed cycle of the fish as well as the corresponding acceleration (Fig. 8.6). As previously shown, the sequence begins with a short acceleration during 2/15 s, and once the fish has reached its maximum speed, it glides and slows down to return to its initial speed (Fig. 8.6c–d). Our robotic fish can mimic relatively well the motion pattern of the zebrafish. The mean speed obtained, as already shown in Fig. 8.4, is similar. The initial speed is smaller in the case of the lure than for the fish; this is explained by the friction of the lure module on the aquarium floor. The standard deviations are relatively similar, which shows that the robotic platform can reproduce well the typical locomotion patterns of a zebrafish.

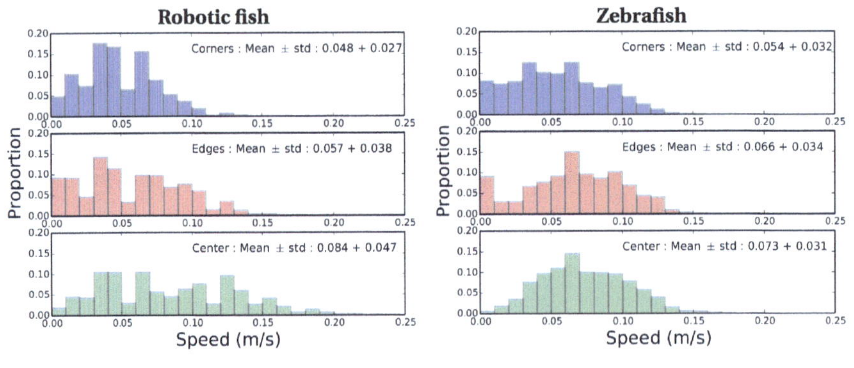

Fig. 8.4 Speed density for the motion of an agent, the RiBot and zebrafish over the span of an hour in the three different zones of the aquarium: corners, edges and center

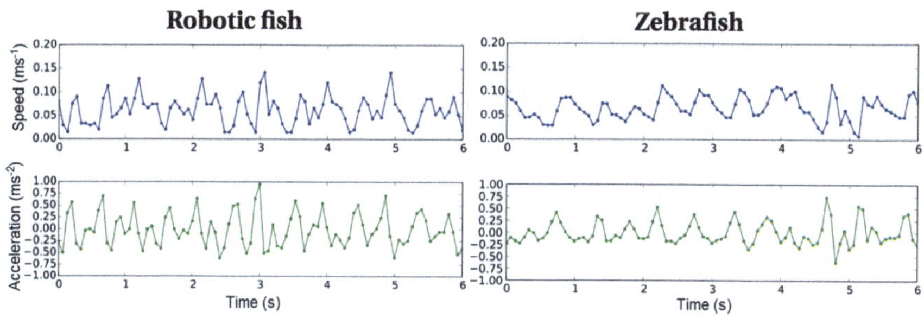

Fig. 8.5 The typical speed and acceleration of an agent, Fish-robot and zebrafish, movements in a corner of the aquarium recorded over 6 s

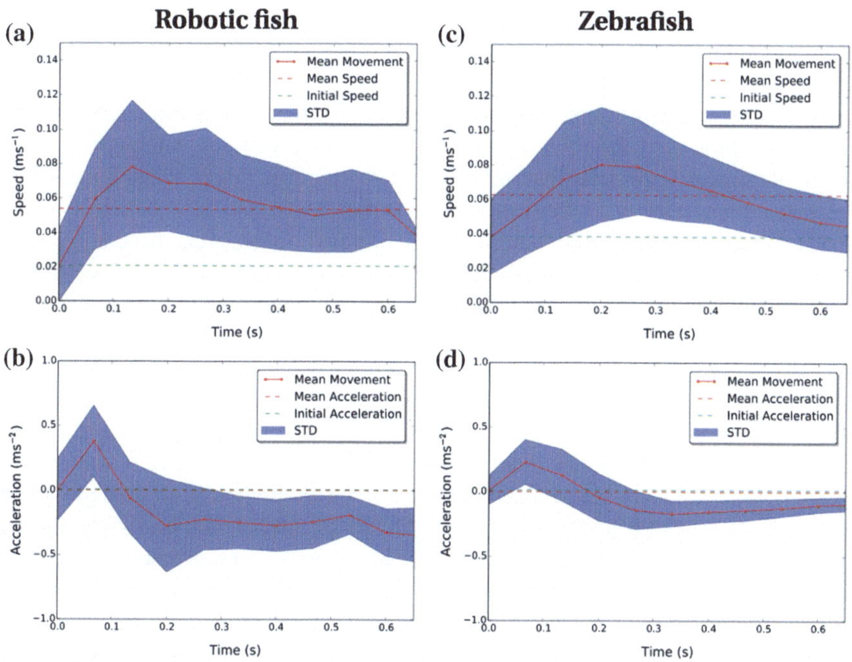

Fig. 8.6 The average agents movement sequence for movement along the edges of the experimental tank. Linear speed (**a**–**c**) and acceleration (**b**–**d**) of the lure and the zebrafish during the speed cycle. The cycle begins with a short acceleration (2/15 s) until the agent reaches its maximum speed, then it decelerates by gliding in the water until the next tail beat. The average was made on several hundreds of samples, determined by the number of sequence made by the agent in one hour

8.3.6 Conclusion

We already showed that the FishBot can achieve very high acceleration and linear speed, similar to the zebrafish ones in Chap. 4. The local obstacle avoidance behavior and the continuous powering of the device allows us to run very long duration experiments without any human intervention during the experiment. Although the RiBot is limited in the range of its tail beating as explained in Chap. 5, it can reproduce the average tail beating range of the zebrafish. Thus the robotic system had the potential to mimic the zebrafish trajectories, locomotion patterns and body movements.

We have studied the locomotion behavior of zebrafish in a square tank in order to establish a controller for the locomotion of the FishBot. First, we observed a heterogeneous spatial repartition of the linear speed of the fish in the open arena. Indeed, the fish tend to swim slowly in the corners and accelerate along the edges and even more so in the center of the tank. Then, the detailed analysis of the motion pattern showed that fish locomotion follows a three-steps sequence: orientation, acceleration and relaxation. First, the zebrafish adjusts its direction. Then, it quickly accelerates in 2/15 s. Finally, it slowly decelerates to return to its initial speed. The succession of these cycles produces a sinusoidal-like evolution of the acceleration and linear speed.

We implemented a finite-state machine with three tunable parameters into the FishBots to reproduce the zebrafish motion sequence underwater. Although there are small differences, especially in terms of speed distribution, this is, to our knowledge, the first locomotion matching between a robotic device and zebrafish that shows that a robot can behave that similarly to a zebrafish. Moreover, the combination of the FishBot locomotion pattern and the body movements of the RiBot offers a wide range of stimuli for behavior research. Indeed, by mimicking the aspect ratio of the zebrafish and being able to reproduce the tail beating frequency of the fish, the RiBot can emit different visual and hydrodynamic cues to interact with fish. The proposed controller also enables the precise and constant adjustment of the acceleration and speed of the FishBot. While the shape and color of the fish lures are important components of their attractiveness (Abaid et al. 2012a), the pattern motion of the robot could also play a key role in the communication with groups of fish. Indeed, the fish perceive and react to the movements of their congeners during collective motion. In particular, rapid changes in orientation or movements performed in front of the group can propagate across the entire school, thanks to the network of visual interaction of the fish (Lemasson et al. 2013; Strandburg-Peshkin et al. 2013; Rosenthal et al. 2015). Therefore, the development of a highly maneuverable robot that is able to reproduce fish locomotion and quickly adjust its trajectory is an important step towards achieving artificial agents that can influence and lead collective motion by emitting similar visual and kinetic signals to the fish.

8.4 Integration of High-Level Controllers to Reproduce the Shoaling Behavior of Zebrafish

We implemented in CATS a vision-based model that was developed by our colleagues from the University Paris Diderot in the project ASSISIbf to generate virtual fish-like trajectories based on the collective behavior of zebrafish in heteroge-

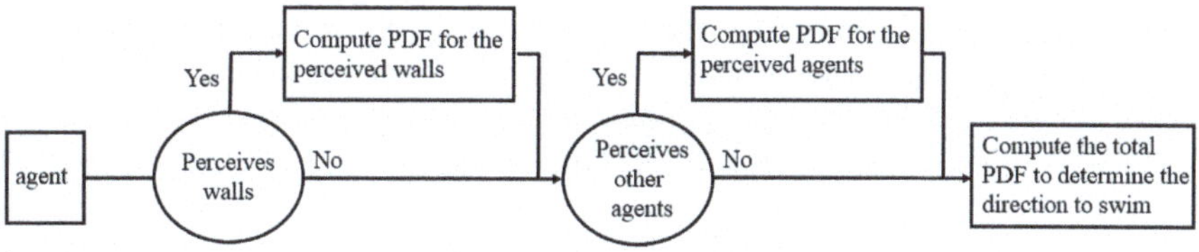

Fig. 8.7 Description of the different steps to compute the orientation Θ_i of a focal fish at each time step using the vision-based model described in detail in Collignon et al. (2016). The proximity to a wall is determined by comparing the distance of the agent to the closest wall with a threshold values $d_w = 15$ cm. The PDF for basic behavior is given by a von Mises distribution centered on 0 while the PDF for wall-following behavior is given by a weighted sum of two von Mises distribution, each of them centered on one of the two possible directions along the wall. Other agents are perceived if they are present in the field of view of the focal agent. The $2nd$ PDF is then computed by weighting von Mises distributions centered on all perceived agents

neous environments (Collignon et al. 2016). The trajectories are injected inside the high-level FishBot controller in CATS to generate the commands to be sent to the Fishbots so that they move to these targets (Figs. 8.7 and 8.8).

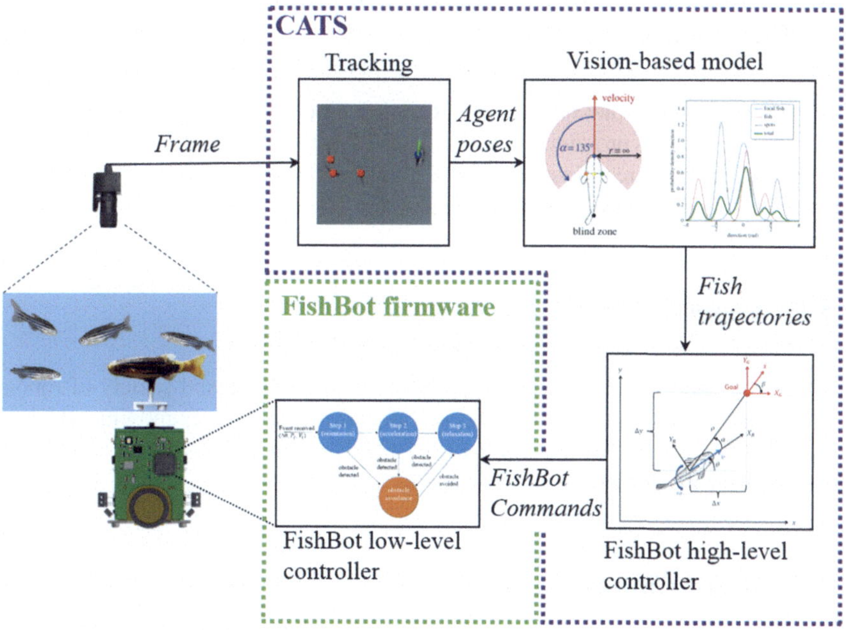

Fig. 8.8 Control architecture to mimic the trajectories and locomotion of zebrafish. The poses of all the agents are retrieved by the visual tracking and used in the vision-based model of Collignon et al. (2016) implemented inside CATS. The estimated agent trajectories are sent to the high-level FishBot controller that is running inside CATS to compute the global command that needs to be sent to the robot to reach the desired target. The commands are received by the FishBot that execute the finite-state machine described in Sect. 8.3.3

In this model, the agents update their position vector X_i with a velocity vector V_i through a discrete time process in a bounded two-dimensional space

$$X_i(t + \delta t) = X_i(t) + V_i(t)\delta t \tag{8.1}$$

$$V_i(t + \delta t) = v_i(t + \delta t)\Theta_i(t + \delta t) \tag{8.2}$$

with v_i the linear speed of the ith agent and Θ_i its orientation. The linear speed v_i of the agent is randomly drawn from the instantaneous speed distribution experimentally measured.

The orientation Θ_i is drawn from a probability density function (PDF) computed as a mixture distribution of von Mises distributions centered on the stimuli perceived by the focal agent. In this study, we only took into account the influence of other agents and the walls of the experimental arena. Thus, this custom PDF is composed by the weighted sum of a PDF taking into account the effect of the walls and a PDF describing the response to other agents. The process is shown in Fig. 8.7.

8.5 Demonstration of the Coupling of High and Low-Level Controllers

Here, we will demonstrate how we could qualified our entire framework composed of the robotic systems presented in Chaps. 4 and 5, our experimental setup presented in Chap. 6, our software infrastructure (Chap. 7) and the controllers developed in this chapter in a single experiment involving a mixed society of robots and zebrafish.

8.5.1 Arena

We used the open circular arena (Fig. 6.5d) to confine the mixed society of robots and animals. In this type of arena, the zebrafish are continuously moving which is not always the case in the open arena (Fig. 6.5a) in which the fish can sometimes aggregate near the corners.

8.5.2 Mixed Group

We have used for this experiment a mixed society of three robots and three zebrafish. The robots were composed of the FishBots steering the RiBots.

The measured positions of the three zebrafish is used as input of the fish collective motion model to generate the trajectories of the three robots. At each control time step, the robots try to reach the targets using the biomimetic locomotion pattern.

8.5.3 Results

First, we have succeeded in controlling multiple robots in a closed-loop experiment for periods of up to 30 min without any human intervention. The closed-loop control was performed in real-time, with robots adapting to the fish behaviors thanks to the vision-based model that could make the robots reproduce the collective movements of the zebrafish.

Figure 8.9, right, presents examples of individual trajectories for zebrafish and robots. The robots can integrate into zebrafish groups that tend to follow the walls of the arena, while mimicking their locomotion patterns. The fish are not afraid by the robotic lure, and we observed the presence of closed-loop interactions: robots tend to follow the fish, and fish can follow a robot.

Figures 8.10 and 8.11 show the individual speed and inter-individual distance for each agent. In Fig. 8.10, we can observe that, in terms of mean and distributions, the robots are able to mimic the zebrafish in terms of speed. However, regarding the inter-individual distances, the fish are able to swim more close than the robot, as it can be observed in Fig. 8.11. The reasons for that is that the vision-based model do not guarantee that the target will always be generated by maintaining a fish-like shoaling distance between the agents. Also, sometimes, near the walls, the robots will have some difficulty to avoid the walls and mimicking the fish behavior that are moving in shoal along the walls. Even if some parameters of the models could be improved, these results are promising, and could be extended to more complex

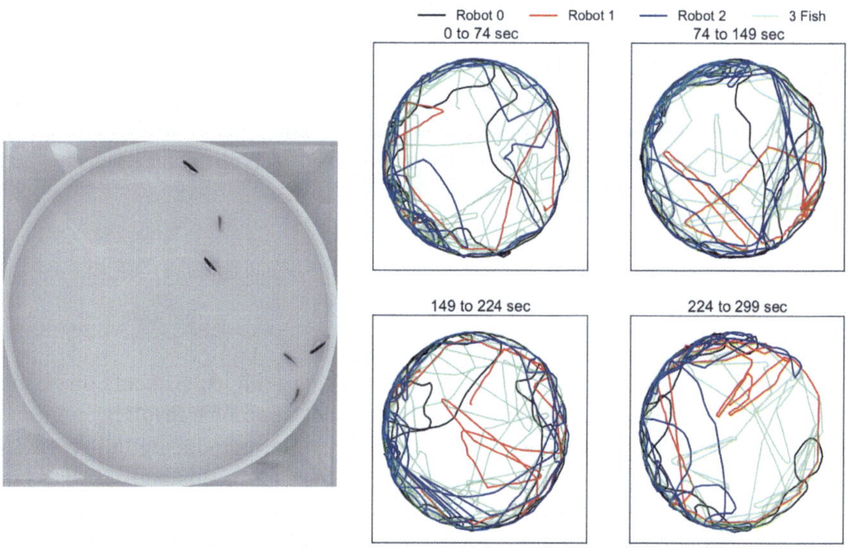

Fig. 8.9 Left: the circular open arena with the mixed group composed of three zebrafish and three robots. Right: Examples of individual trajectories from fish and robots in a circular arena. We used three fish-robots and three zebrafish. 300 s (4500 frames) of one experiment are represented

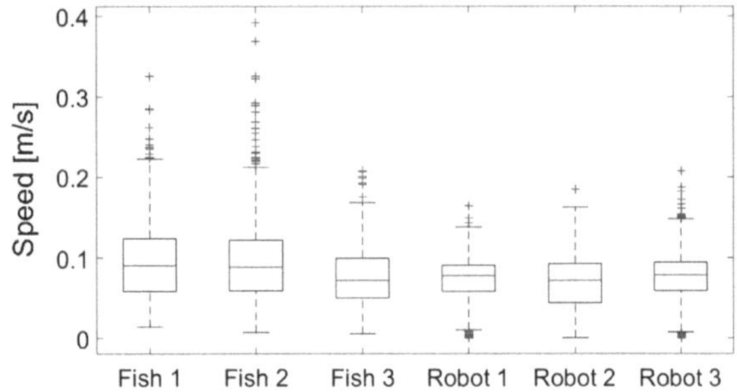

Fig. 8.10 Agents' speed in the mixed group experiment in an open circular arena. The individual speed was extracted using idTracker (Pérez-Escudero et al. 2014). 300 s (4500 frames) of one experiment are represented

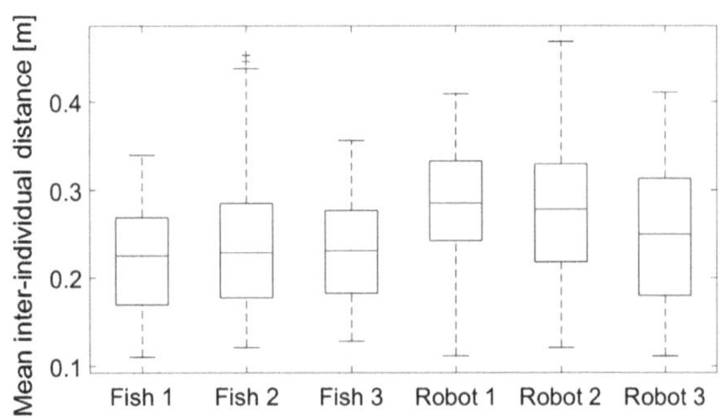

Fig. 8.11 Inter-individual distance between agents in the mixed group experiment in an open circular arena. The individual position at each time step was extracted using idTracker (Pérez-Escudero et al. 2014). 300 s (4500 frames) of one experiment are represented

setups, with larger populations of fish and robots. For instance, one of the setup that will be used in the context of the ASSISIbf project is the setup composed of two squared arenas separated by a corridor (Fig. 6.5b). This setup is convenient as it offers a binary choice where the agents spend most of the time in the two rooms, transiting fast between the two rooms. The behavior of the zebrafish inside the rooms is similar to the one inside the open circular arena. Thus, the controllers designed in this chapter are currently applied in mixed groups of zebrafish and robots in such arenas in the context of the project ASSISIbf.

8.6 Contributions to the State of the Art

As we shown in Chap. 2, most of the groups working with experiments involving mixed groups of fish and robots have not yet shown an experiment involving multiple robots mimicking the trajectories of the zebrafish while reproducing the zebrafish locomotion patterns. We are thus the first group to achieve this and validate our controllers in experiments involving mixed groups of fish and robots. We hope that the system designed will help biologists for further investigating the importance of biomimetic controllers when building mixed societies of fish and robots.

8.7 People Who Contributed to This Work

Axel Séguret made the experiment with zebrafish in the aquarium in order to analyze the characteristics of their swimming patterns. Dr. Bertrand Collignon performed part of the analysis to extract the characteristics of the zebrafish swimming patterns. Léo Cazenille implemented the vision-based model so that it can be used in CATS.

Chapter 9
Application of a Fractional Factorial Design to Model the Attractiveness of a Robotic Fish to a Shoal of Zebrafish

9.1 Summary

To design the appropriate robotic system used during mixed society experiments, a stimuli analysis was preliminary required to test which aspects of the robotic devices could trigger the social behavior of the zebrafish. The main difficulty with such stimuli analysis is to isolate the effect of the different stimuli that the robot generates and that could induce a social behavior on the fish. In order to study the acceptance of the robot towards zebrafish, we varied several parameters of the system and used design of experiments methods to reduce the number of performed experiments and determined the impact of each factor on the acceptance of the robot towards a small group of zebrafish (Fig. 9.1).

This chapter is based on the publication *A Miniature Mobile Robot Developed to be Socially Integrated with Species of Small Fish* (Bonnet et al. 2014) with slight modifications to fit the present dissertation.

9.2 Description of the Experiment

This chapter as well as the two next chapters will mainly describe experiments performed with zebrafish and robots, in which the behavior of the zebrafish was analyzed depending on different stimuli that the robot can generate.

In this chapter in particular, we will describe an experiment that was conducted during the design phase of the FishBot. At that time, its main purpose was to assess which configuration of the robot design and control needed to be improved to increase the acceptance of the robots to the zebrafish. These tests were necessary in our quest to build a mixed society of fish and robots, as defined in Sect. 1.2.3. During this phase, we also wanted to be sure that our robotic system could attract the fish and not repel them. One major difficulty of these types of tests is the large number of

F. Bonnet and F. Mondada, *Shoaling with Fish: Using Miniature Robotic Agents to Close the Interaction Loop with Groups of Zebrafish Danio rerio*, Springer Tracts in Advanced Robotics 131, https://doi.org/10.1007/978-3-030-16781-3_9

Fig. 9.1 The FishBot coupled with a fishing lure that is moving inside an aquarium among zebrafish

factors that need to be tested in order to measure their effect on the robot acceptance. We selected four of these factors that we supposed having the most effect: the linear speed of the robot, the visual appearance of the lure, the type of trajectory made by the robot and the motion continuity of the lure. We used a fractional factorial design (Hunter et al. 1978) in order to optimize the number of experiments and to determine the acceptance of the robot by the zebrafish with respect to the different parameters.

9.2.1 Experimental Setup

The experimental arena that was used for this experiment consisted of the open arena presented in Sect. 6.3 (Fig. 6.5a). The zebrafish were thus free to move everywhere in the aquarium. One FishBot V3.0 coupled with two types of lures was used to perform this experiment. In order to control the FishBot in a closed-loop, we used the initial tracking and control software made in Python language (Sect. 7.3.1).

9.2.2 Lure Module

Two different lures were used (Fig. 9.2). The first one is a 3D printed rigid black ellipsoid made of Acrylonitrile Butadiene Styrene (ABS) with a length of 40 mm, width and height of 5 mm, and the second one is a soft fishing lure bought in a shop (Decathlon, France) with a length of 45 mm. We chose to test two lures with approximately the same size, but one without any biomimetic cues and one with multiple biomimetic cues in order to determine if the shape of the conspecifics plays a major role for the zebrafish.

The lure was moving at a constant height which depended on the height of the support. In this case, we chose a support's height of 5 cm in order to have the lure moving at the middle of the water level which was of 10 cm at that time. The base of the lure module was painted in white in order to blend into the white background. The lure was glued on the base and could not rotate.

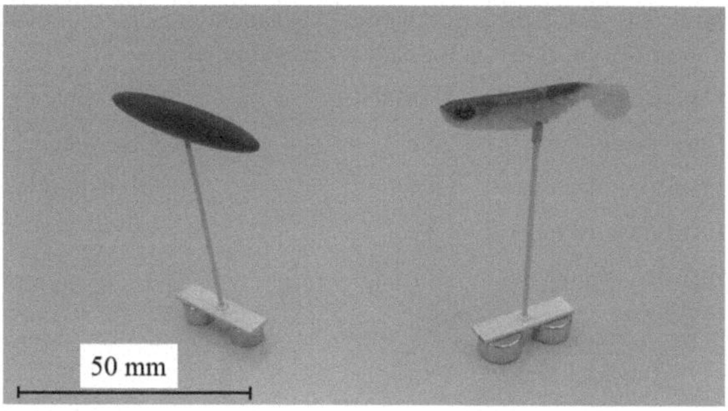

Fig. 9.2 Lures that were used to perform the experiments involving the FishBot V3.0. Left: a rigid black ellipsoid lure. Right: a soft fishing lure. The base was painted in white so that it is merged with the white background inside the aquarium

9.2.3 Experimental Procedure

For each experiment, three zebrafish out of a shoal of 20 individuals were placed inside the experimental tank. The three zebrafish were selected randomly among the entire group in order to reduce the learning possibility by the fish from the previous experiments. For the rest, we used the same experimental procedure than the one presented in Sect. 6.4, and the duration of the experiments was 30 min long.

9.2.4 Factors and Measurements

In this experiment, four factors were considered (Table 9.1): the lure shape, the Fishbot trajectory, movement and speed. The first studied factor was the shape of the lure: it can either be of ellipsoidal shape or represent a fake fish (Fig. 9.2). The idea here is to assess if the visual appearance of the lure has an influence on the perception of the zebrafish. The second factor describes the different possible trajectories of the robot: it can either turn in circles in the whole aquarium or alternate between full circles of diameter of 60 cm, and alternate between full and half-circles (Fig. 9.3). Indeed, the zebrafish have a tendency to swim along the border of the tank, as they feel more protected near walls than in the central area of the tank (Reed and Jennings 2011). We wanted to analyze if the robotic fish could influence the group of zebrafish to explore the central area. The FishBot movements was also included as a third factor: the robot could either move continuously or with forward jerks (thus mimicking more closely the fish behavior). Finally, the robot speed was a continuous factor, varying between 0.03 and 0.06 $\mathrm{m\,s^{-1}}$.

We measured the attractiveness of the robot using two parameters. The first response analyzed was the mean distance between the zebrafish and the lure during the experiments. By measuring this value, one can assess the acceptance of the robotic fish in the fish shoal. It was measured in millimeters. The second analyzed response was the number of times during the experiment a zebrafish was located

Table 9.1 The four factors of the experiment presented in this chapter involving the third version of the FishBot. Each factor had two possible values

Factor name	Variable	Min value (−1)	Max value (+1)
Lure shape	X_1	Fake fish	Ellipsoid
FishBot trajectory	X_2	(Half-)circles	Circles
FishBot movement	X_3	Jerks	Continuous
FishBot speed ($\mathrm{cm\,s^{-1}}$)	X_4	3	6

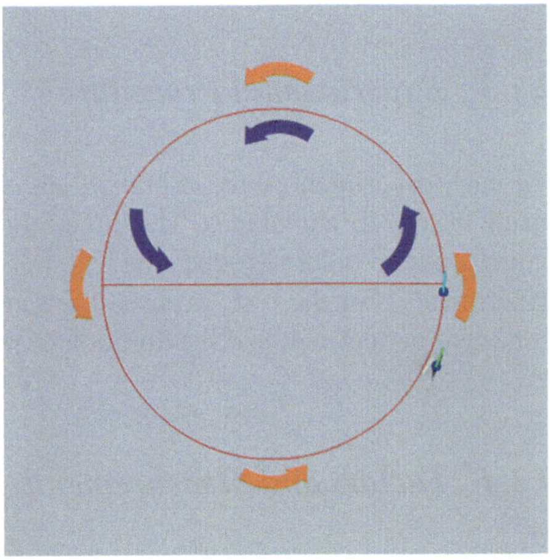

Fig. 9.3 The two trajectories (in red) followed by the robotic fish (blue dot with the green line). In one of the two programmed trajectory, the robot is only rotating in circle (orange arrows). In the other one, the robot is half of the time following the circle (orange arrows) and half of the time moving towards the center of the aquarium, thus crossing the circle (purple arrows)

inside a radius of 10 cm around the lure, thus approximately two zebrafish body lengths.

At that time of the project, due to the fact that we only had 20 zebrafish and that we were making preliminary experiments, we considered that one trial per experiment was relevant enough for the tests that we had to perform, as each factor was tested in four conditions. However, we noticed that there was a large variability in the result, due to the fact that the behavior of the zebrafish could vary between experiments. Therefore, in the experiments that were performed later in the project and which results will be described in the next two chapters, we performed more trials per condition in order to obtain a more robust analysis.

9.3 Design of Experiment

A linear model with interactions was selected to analyze the effect of the four different factors on the response of the zebrafish.

The model can be written as follows (Montgomery 2008):

$$Y = a_0 + \sum_{i=1}^{n} a_i X_i + \sum_{i,j=1,j \neq i}^{n} a_{ij} X_i X_j, \tag{9.1}$$

where the X_i stand for the values of the factors, a_i are the model coefficients associated to X_i, Y is the response variable and n the number of factor, four in the case of this study.

In a full factorial design, a total of $2^4 = 16$ experiments would have been needed, due to the fact that we had four factors with two possible conditions. In order to reduce the number of experiments (which are time- and resource-consuming), a fractional factorial design of type 2_{IV}^{4-1} was used (Hunter et al. 1978). Hence, with such design, only $2^3 = 8$ experiments were needed. This reduction in the number of experiments comes at a cost, however, some of the coefficients were then aliased. This design having a resolution of IV, neither main effects a_i are confounded between each other, nor main effects with first-level interaction terms a_{ij} (Hunter et al. 1978). However, first-level interaction terms a_{ij} are confounded between each other and main effects a_i are confounded with second-level interaction coefficients a_{ijk} (Furbringer 2005). If one neglects second-level and higher-level interaction terms, this design is able to determine the main effects without bias (Montgomery 2008). However, the first-level interaction terms are aliased: a_{12} with a_{34}, a_{13} with a_{24} and a_{14} with a_{23}.

The following matrix of experiments E for the 2_{IV}^{4-1} fractional factorial design was obtained:

$$E = \begin{array}{c} \begin{matrix} X_1 & X_2 & X_3 & X_4 \end{matrix} \\ \begin{bmatrix} 1 & 1 & 1 & 1 \\ 1 & 1 & -1 & -1 \\ 1 & -1 & 1 & -1 \\ 1 & -1 & -1 & 1 \\ -1 & 1 & 1 & -1 \\ -1 & 1 & -1 & 1 \\ -1 & -1 & 1 & 1 \\ -1 & -1 & -1 & -1 \end{bmatrix} \end{array}$$

Hence, the matrix of the model with interactions X was given by

$$X = \begin{array}{c} \begin{matrix} a_0 & X_1 & X_2 & X_3 & X_4 & X_{12} & X_{13} & X_{14} \end{matrix} \\ \begin{bmatrix} 1 & 1 & 1 & 1 & 1 & 1 & 1 & 1 \\ 1 & 1 & 1 & -1 & -1 & 1 & -1 & -1 \\ 1 & 1 & -1 & 1 & -1 & -1 & 1 & -1 \\ 1 & 1 & -1 & -1 & 1 & -1 & -1 & 1 \\ 1 & -1 & 1 & 1 & -1 & -1 & -1 & 1 \\ 1 & -1 & 1 & -1 & 1 & -1 & 1 & -1 \\ 1 & -1 & -1 & 1 & 1 & 1 & -1 & -1 \\ 1 & -1 & -1 & -1 & -1 & 1 & 1 & 1 \end{bmatrix} \end{array}$$

where the first column corresponds to the constant coefficient a_0 and is a column of '1', the next four columns correspond to the main effects a_1, a_2, a_3 and a_4 and are composed of the four columns of the matrix of experiments E. The last three columns were obtained by multiplying the columns of the matrix of experiments as they correspond to the interacting factors (Furbringer 2005).

9.4 Results and Discussion

Table 9.2 shows the results that were obtained from the eight performed experiments.

In order to assess if the fish seem attracted by the lure, we have compared the results obtained from the mean distances with an experiment in which only the three zebrafish were swimming, without any lure moving inside the tank. We have simulated a virtual lure that reproduced the same movements as the robot and computed the distance between each zebrafish and this virtual lure. We have obtained a value of 465.88 mm for the average distance, which is above all the distances obtained during the experiments (Table 9.2). Thus, we had an indication that the zebrafish seem to be more convinced to explore the center of the aquarium and swim close to the lure than when no objects are moving inside of the tank, showing that the robotic fish had no repulsive effects. However, when measuring the percentage of presence of zebrafish in the zone of 10 cm of radius around the lure, we obtain percentages of less than 10% of the time that one zebrafish was near the lure, meaning that, most of the time, the zebrafish were swimming in shoal near the walls of the arena and ignoring the lure.

The coefficients of the linear model have been estimated using a least squares regression:

$$\hat{a} = (X^T X)^{-1} X^T Y, \tag{9.2}$$

where Y is the vector of results. As the factorial design is orthogonal (Hunter et al. 1978), this equation reduces to

Table 9.2 Results obtained from the eight experiments performed. One trial of 30 min was performed for each condition. We measured two parameters: the mean distance between the zebrafish and the lure $\bar{d}$ and the number of time a zebrafish was inside a radius of 10 cm from the lure T

Run	Mean distance ($\bar{d}$ (mm))	# of times near robot (T (-))
1	414.27	274
2	354.05	2895
3	420.90	1003
4	372.99	1706
5	401.88	1327
6	407.87	1173
7	407.57	949
8	390.47	1325

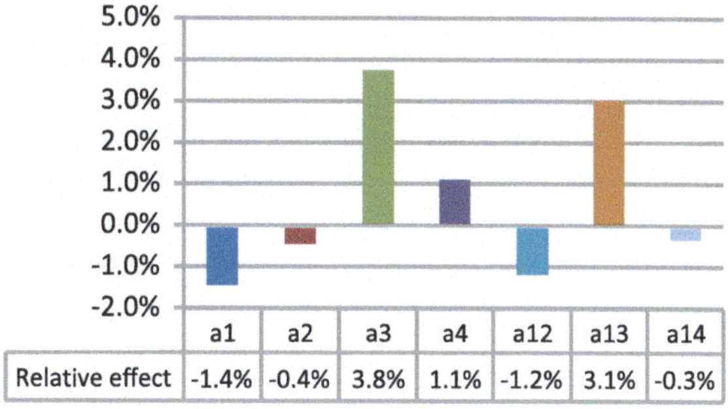

Fig. 9.4 Relative effects of the factors on the first measure, the mean distance between the lure and the zebrafish ($\bar{d}$)

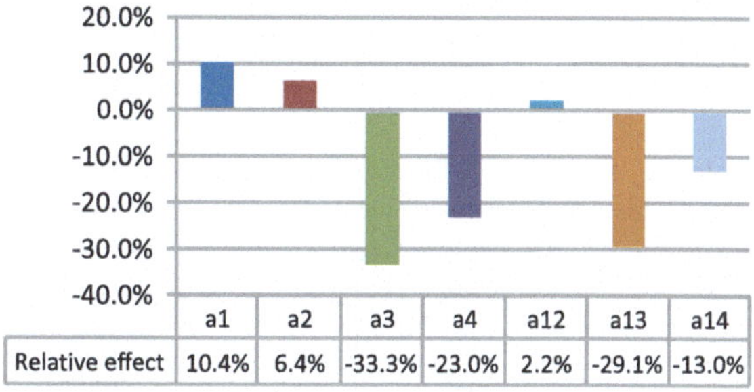

Fig. 9.5 Relative effects of the factors on the second measure, the number of times a zebrafish was situated at less than 10 cm from the lure (T)

$$\hat{a} = \frac{1}{N} X^T Y, \tag{9.3}$$

where N is the number of experiments ($N = 8$ in our case).

Figure 9.4 shows the relative effects of the different factors on the mean distance ($\bar{d}$) between the lure and the zebrafish. It allows to obtain a first understanding about the most influential factors on the response. The robot's movements (a_3) and the interaction terms between the lure shape and the robot's movements (a_{13}) seem to be the most influential factors. However, the effects are limited: none of the factors had a relative effect above 5%.

Figure 9.5 shows the relative effects of the different factors on the number of times (T) a zebrafish is situated at less than 10 cm of the lure. The same initial conclusions can be drawn than for $\bar{d}$: the most influential factors on the response are the robot's movement (a_3) and the interaction term between the shape of the lure and the movements of the FishBot (a_{13}), followed by the robot's speed (a_4). Furthermore, all coefficients (except a_{12}) have a relative effect above 5%.

These results are coherent as we can observe that in the case of $\bar{d}$, a small response value indicates a higher acceptation rate of the lure (because of a smaller mean distance), whereas in the case of T, a high acceptation rate of the lure is indicated by an elevated response value. As the relative effects of the coefficients were more significant for T, we decided to investigate further the results concerning this measure.

Table 9.3 presents the ANOVA table of response T. In a first step, we decided to not include factors with a relative effect below 5% in the residual (only a_{12} in this

Table 9.3 ANOVA table for response T, where factors with relative effects below 5% are included in the residual

Effect	SS	df	MS	F	p
a_0	14183138	1	14183138	2072.2	0.014
a_1	152352	1	152352	22.3	0.133
a_2	58825	1	58825	8.6	0.209
a_3	1571765	1	1571765	229.6	0.042
a_4	749088	1	749088	109.4	0.061
a_{13}	1202801	1	1202801	175.7	0.048
a_{14}	240818	1	240818	35.2	0.106
Residual	6845	1	6845		
Total	18165630	8			

Table 9.4 ANOVA table for response T, where all factors except the three main factors and the constant are included in the residual

Effect	SS	df	MS	F	p
a_0	14183138	1	14183138	123.6	0.571
a_3	1571765	1	1571765	13.7	0.168
a_4	749088	1	749088	6.5	0.237
a_{13}	1202801	1	1202801	10.5	0.191
Residual	458839	4	114710		
Total	18165630	8			

case). The ANOVA results show that the three main effects described earlier (a_3, a_{13} and a_4) have a probability of being random of 4.2%, 4.8% and 6.1% respectively, and can thus be considered as significant effects. The effects of the other factors are not certain because their probabilities of being random are much higher (between 10 and 20%). This uncertainty on the other effects is mainly due to the low number of experiments associated with a fractional factorial design (compared to the number of coefficients), which in turn decreases the degrees of freedom left for the residual. One solution would be to reduce the number of coefficients in order to increase the degrees of freedom left for the residual. However, as Table 9.4 shows, including all factors except the three main factors and the constant in the residual does not improve the p-value of the remaining factors. Finally, it was decided to keep all factors with a relative effect above 5% in the linear model presented in Eq. (9.1). Thus the following model for the response T was obtained:

$$Y_T = \sum_{i=1}^{4} a_i X_i + a_{13} X_{13} + a_{14} X_{14} + Res, \tag{9.4}$$

where Y_T is the vector of results obtained for the response T and Res the residual.

9.5 Conclusion

We investigated the acceptance of a robotic fish among a group of real zebrafish while varying several parameters of the experiments, such as the shape of the lure and the movement of the robot in the tank.

We noticed that the zebrafish seem to be attracted by the lure, by comparing experiments with and without the lure moving inside the tank. However, the fish were swimming less than 10% of the time near the lure during the 30 min long experiments. We realized that the open area setup was not constraining enough the zebrafish movements, as the zebrafish preferred to explore the arena than swim in the direction of the lure. Therefore, we decided at that time of the project to create new types of experimental arenas that could constrain the fish movements so that they would swim more close to the robotic agents, and also it would not be too complex for the robots to join the shoal.

Statistical analysis were performed to process the data and build a model. We used a fractional factorial design to reduce the number of experiments to be performed. The reduction of experiments is also a powerful tool when working with real animals, as it reduces also the amount of required subjects.

Two measurements were considered: the mean distances between the fish and the robot, and the number of time that a zebrafish was at less than 10 cm from the robot during the experiment, thus inside a zone in which shoaling can be considered. As the relative effects of the coefficients were more significant for the latter, we decided to investigate further the results concerning the second measure.

Results showed that among the different parameters that were varied during the experiments, coefficients corresponding respectively to the robot's movement, speed and interaction between the mock-up shape and movements of the robot could be considered as being significant, and thus included in the model. As the p-value was not improved by reducing the number of factors, it was decided to keep all the factors in the linear model except the interaction term between the shape of the lure and the robot trajectory.

To conclude, we can state that the locomotion of the lure is crucial for its acceptance towards the animal. We have also shown that the visual appearance of the lure has to be taken into account, thus, at that time of the project, we decided to create a biomimetic robotic lure, the RiBot, described in Chap. 5, to investigate more deeply the effect of the lure on its acceptance towards zebrafish. A more detailed study on the effect of the lure on the collective decisions of zebrafish will be described in the next chapter.

9.6 People Who Contributed to This Work

Stefan Binder contributed to the design of experiment method research and to the statistical analysis.

Chapter 10
Using a Circular Corridor to Characterize the Attractive Cues of Lures for a Shoal of Zebrafish

10.1 Summary

The previous chapter presented the results of an experiment performed in order to isolate and model the effect of various stimuli that the robotic device could generate on its attractiveness to a fish shoal. We showed that, mostly, the robot's movements and the lure shape had an effect on the acceptance of the robot. However, the effect of the lure shape was not analyzed in detail.

In order to determine the most attractive lure for this type of experiment involving the FishBot, we designed an experimental setup composed of a circular corridor and a coaxial motor that can steer lures inside the corridor. The attractiveness of the lures can be measured using the decision of the fish to swim in the same direction as the lure (Fig. 10.1). In this chapter, we present the results of three experiments comparing the attractiveness of different lures on a shoal of zebrafish.

10.2 Methods

10.2.1 Experimental Setup

The experimental setup designed for this study is presented in Fig. 10.2 left. We replaced the support of the FishBot (Fig. 6.4f) underneath the tank with a coaxial motor rotating two rotors in the two possible directions, clockwise (CW) or counter-clockwise (CCW). Magnets are placed at the tip of these rotors in order to transmit the motion to lures that are moving inside the tank.

In the experiments that are performed in this study, the lures are controlled in an open-loop. This offers the advantage that we can test any type of lure, and the lures do not have to be tracked in order to control them as in a closed-loop.

F. Bonnet and F. Mondada, *Shoaling with Fish: Using Miniature Robotic Agents to Close the Interaction Loop with Groups of Zebrafish Danio rerio*, Springer Tracts in Advanced Robotics 131, https://doi.org/10.1007/978-3-030-16781-3_10

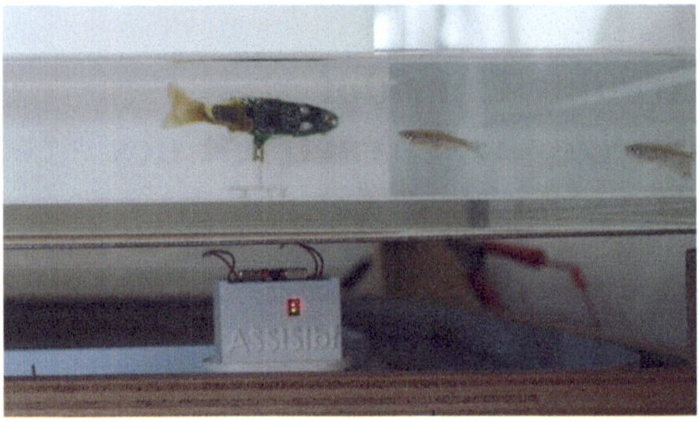

Fig. 10.1 The RiBot facing a zebrafish

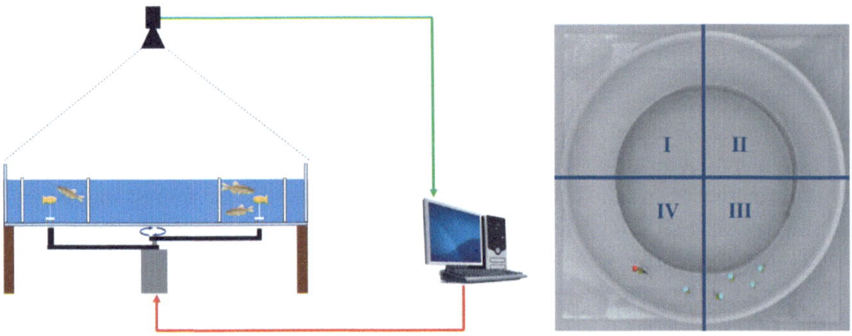

Fig. 10.2 Left: experimental setup used for the experiments described in this chapter. A coaxial motor is rotating two blades underneath the experimental tank presented in Chap. 6. The two blades can rotate in both directions independently and in a different range of speeds that can be controlled from a computer. Inside the tank, the fish and the lures are constrained in a circular corridor. A camera is placed on top of the tank and grabs frames that are collected on the computer that performs the online and offline tracking of the agents. Right: The result of the online tracking performed on the low-resolution frames grabbed by the camera placed on top of the experimental setup. The position of the five zebrafish (cyan dots) and the lure (red dot) is retrieved and the swimming direction of the fish is estimated using the position of the fish groups. The setup is cut into four quadrants during the analysis in order to estimate the direction in which the fish are swimming. The number of fish inside each quadrant is compared between the frames in order to determine the direction in which the majority of fish is moving

In order to constrain the zebrafish, the arena is composed of an outer circular wall and an inner circular wall that forms a circular corridor inside the tank (Fig. 10.2) that was already described in Fig. 6.5c. The dimensions of the corridor were as follows: an external diameter of 58 cm, an internal diameter of 38 cm, and thus the width of the corridor is 10 cm. The choice of the 10 cm width is a good tradeoff, allowing the zebrafish to have continuous motion without being stressed. Indeed, in a large area, the zebrafish will tend to either move along the walls or stay in one place due to stress, but, in a very narrow corridor, they will move faster as they are stressed by the lack of room. The choice of the outer diameter and the size of the corridor was determined by the non-visibility of the opposite side of the corridor and the fact that with a linear speed of 12 cm s^{-1}, if we suppose that the fish are not moving faster than on average in one direction, it gives the lure time to complete one turn in the setup so that the fish will spend several seconds without the presence of the lure.

10.2.2 Lures

The lures are moving at a constant water level, which depends on the height of the support. We chose a support height of 3 cm in order to have the lure moving at the middle of the water level. The lures were all placed on the module composed of a carbon pin attached to an iron plate on which two magnets were placed. The module was painted in white so that it blends in with the white background of the tank.

The linear speed of the lure for each experiment was 12 cm s^{-1}. This speed is higher than the average speed of the fish in the circular corridor (8 cm s^{-1}), as we wanted to increase the attractive effect of the lure. Indeed, when escaping from the group and possibly attracting the rest of the shoal with them, the fish will have a tendency to increase their linear speed, and thus a robot with a linear speed higher than the zebrafish shoal could have a higher chance of attracting the zebrafish.

The goal of this study was to test various types of lures, which will be described in detail in the different experiments described in this chapter.

10.2.3 Tracking

In order to determine the swimming direction of the fish, we separated the experimental setup into four quadrants (Fig. 10.2 right). Each second, the algorithm finds the quadrant in which the majority of fish is currently swimming. Then it is compared with the result in the next frame. It will be counted as turning CW or CCW depending on the new estimated position of the fish shoal. The percentage of swimming direction is obtained using only the cases when the fish are turning, meaning that we ignore all the cases in which the fish are not counted as turning in the setup, which also appends if there is no majority measured (example 2-2-1-0).

In addition to the online tracking, we used high-resolution videos, which we processed offline using idTracker software (Pérez-Escudero et al. 2014). We used idTracker to process some of the experiments in order to first evaluate the algorithm that is used for the online tracking to measure the fish swimming direction, and, second to estimate the distance between the zebrafish and the lure as well as the speed of the zebrafish during the experiment. This allows us to measure whether the zebrafish are stressed during the experiment by comparing their individual linear speeds between the control experiments and the experiment involving lures, and by determining if the fish are swimming close to the robot or not, as they could also swim in the same direction as the robot if they were frightened by it.

Table 10.1 Qualitative effect of the shoal size on different parameters involved in the measurements of the collective behavior of the zebrafish in experiments involving the circular corridor (++: optimal result, +: good result, −: poor result, −−: very poor result)

Number of fish	Group's homogeneity	Attraction effect on the group	Risk of freezing
1	++	−−	−−
2	++	−	−
5	+	+	++
8	−	++	++
10	−−	++	++

10.2.4 *Zebrafish*

For the experiments performed in this chapter, we used 60 wild-type zebrafish, with short fins. We used preliminary experiments to determine the optimal zebrafish shoal size required in order to obtain a clear collective decision response of the shoal. This was needed to quantify the effect of the robotic agents on the collective behavior of the zebrafish. Table 10.1 shows a comparative qualitative study of the different shoal size that were tested in preliminary experiments. With a shoal size smaller than five individuals, the group is very homogenous, but due to the low number of fish, there is a lower chance that one of them will be attracted to the robot, which lowers the chance that an effect will be seen on the whole group. There is also a high probability that the zebrafish will freeze, probably due to the fact that they usually live with many other fish in their housing aquarium and feel stressed when there are only a few conspecifics in an open area. For the case of shoals composed of more than five fish, we observed that the effect of the robot on individuals increases, although the homogeneity of the group will decrease with the increase in the zebrafish number. We observed that the best trade-off between the observed effect and the homogeneity was for groups of five to six zebrafish, and therefore we only performed the experiments presented in this study only with groups of five zebrafish with one lure.

10.2.5 *Hypothesis Tested in Our Experiment*

In all experiments presented in this chapter we made a general hypothesis to determine whether or not the lure has an influence on the collective decision of the fish. Hypothesis H_1 in our experiment is that the lure has a positive influence on the fish shoal's swimming direction, i.e., the fish shoal will tend to swim in the same direction as the lure. Thus, the null hypothesis H_0 is that the lure has no influence on the fish shoal direction of swimming.

10.3 Experiments 0: Testing the Influence of a Constant Water Flux Inside the Circular Corridor

It is known that the behavior of zebrafish might change depending on the presence of water flow (Suriyampola et al. 2016). The flow might, for instance, change their average swimming speed and group cohesion. When an object is rotating inside the circular corridor that we designed, with a constant speed, a water flow is generated after a certain amount of time. This can thus result in a change in the behavior of the zebrafish. Before conducting any test involving different types of lures in the setup, we wanted to analyze the impact of a constant water current inside the circular corridor setup on the behavior of the zebrafish.

10.3.1 Experimental Design

We used an external water filter Eheim Echo Pro 130 (EHEIM GmbH & Co KG, Germany) to generate a water current inside the circular corridor. A pipe connected to the output of the filter was placed inside the corridor and generated a water flow in one direction (Fig. 10.3). We generated a water flow higher than the one created by a lure moving with a constant speed of $12\,\mathrm{cm\,s^{-1}}$, and we used a small floating object in order to determine this speed difference. We performed experiments with 20 min in duration, as we noticed that after 20 min, the zebrafish were getting tired due to the fact that they were constantly swimming against the current, as it will be shown, and we did not want to take risks.

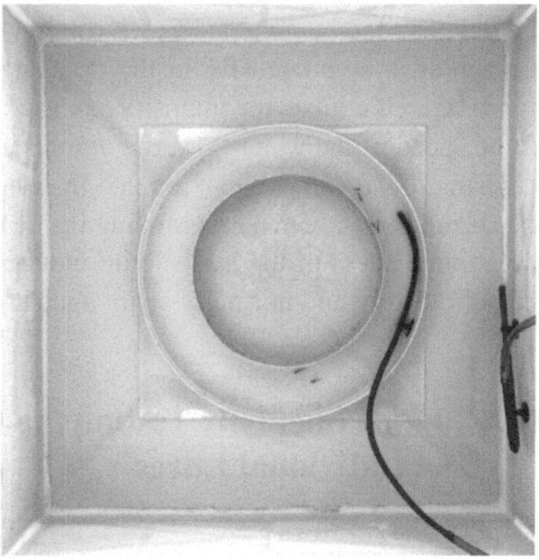

Fig. 10.3 Setup used to test the effect of water currents on the zebrafish behavior. A pipe connected to the output of an external water filter was placed inside the circular corridor in order to generate a water flow in one direction. A shoal of six zebrafish was then placed inside the corridor and we measured their swimming direction in order to determine if they swam mostly with or against the water current

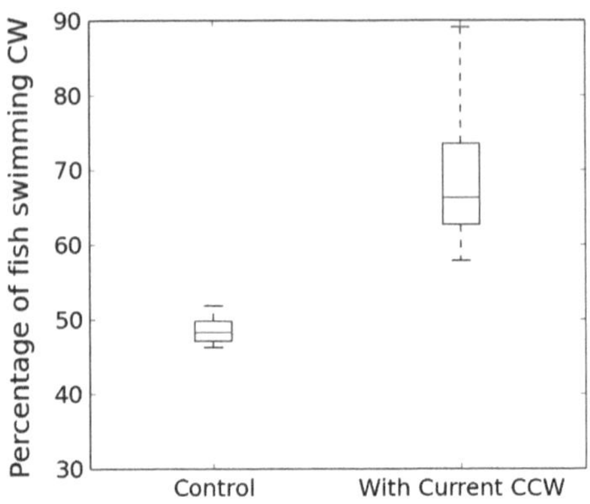

Fig. 10.4 Zebrafish shoal swimming direction when a water current is created inside the circular corridor in one direction (CCW). Each condition was repeated four times using a shoal of six zebrafish that were randomly selected from their housing aquarium. The duration of the experiment was 20 min. The fish that were used for an experiment were not reused for the same experiment afterwards. The distributions differ significantly (Kruskal–Wallis $p < 0.05$), showing that the current has an impact on the zebrafish direction of swimming

10.3.2 Results

First, we could clearly observe that the fish had a tendency to swim against the water current, for any current generated inside the experimental setup. We also observed that for a very strong current, the fish swam strongly in the opposite direction, and thus we avoided long experiments since it seemed they were not healthy for the zebrafish and they were not relevant for what we wanted to show in this experiment.

Figure 10.4 shows the result of the experiment with a water current stronger than the one that can be generated by a robot moving at $12\,\mathrm{cm\,s^{-1}}$. We can observe that the zebrafish swim mostly in the opposite direction of the flow (Kruskal–Wallis $p < 0.05$). Of course, the intensity can vary depending on different parameters, such as the speed of the water flow, the number of fish, the size of the fish, etc.

Using this result, we can assume that the fish will not have a tendency to follow the current; thus, with a lure moving inside the corridor and generating a water flow, the attractiveness of the lure should induce a change in the swimming direction, as the zebrafish would normally swim against the water flow.

10.4 Experiments 1: Testing the Appearance and Shape of Individual Lures

The first parameters that were tested were the visual appearance and shape of the lure. These parameters were hard to decouple, and therefore we combined them in the same experiment. This was used to determine if designing lures that mimic the visual appearance and shape of the zebrafish was necessary to design a lure suited for experiments involving mixed societies of fish and robots.

10.4.1 Experimental Design

Five types of conditions were tested using four different lures (Fig. 10.5):

- Control: The coaxial motor is rotating, CW or CCW, but no lures are placed inside the corridor, and thus nothing is visible to the zebrafish. This condition was tested in order to determine whether or not the coaxial motor had an effect on the swimming direction of the zebrafish.
- Lure base: Only the base of the lure that is painted white is placed without any lure mounted on it. We wanted to determine if the base of the lure would have an attractive or a repulsive effect on the fish.
- 3D lure: A black 3D printed lure that is based on a 3D scan of one of our zebrafish is placed on the lure base. This was used to test if the shape of the lure, without any other visual cues, could be attractive. The fact that this lure has exactly the body shape of a zebrafish is also a good indicator to see if the drag generated by the lure moving could have an attractive effect on the zebrafish shoal.
- Fishing lure: A fishing lure with a size and shape similar to that of the zebrafish, with the tail beating passively when moving. This was used to test if a lure that

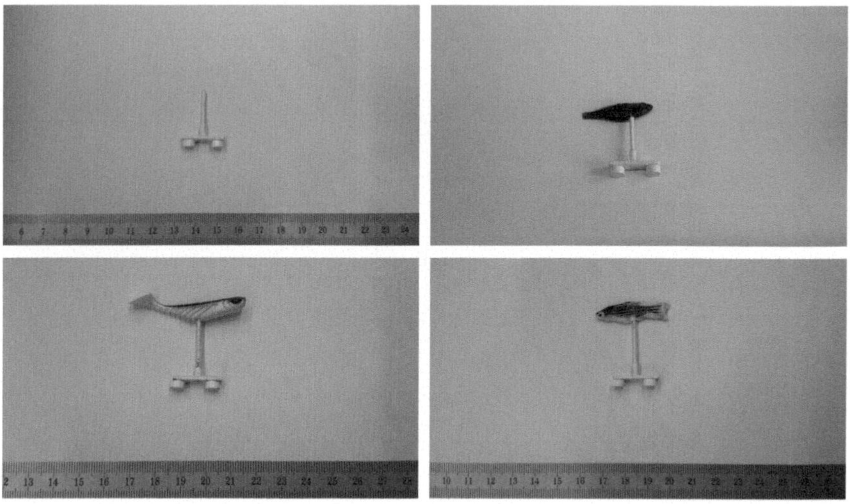

Fig. 10.5 The four lures that were tested in Experiment 1. Top left: the base of the lure composed of an iron plate and a carbon stick of 1.5 mm in diameter painted in white. Top right: a 3D printed zebrafish replica made of ABS fixed on the base. Bottom left: a fishing lure mimicking the fish size aspect ratio and beating its tail passively during the motion of the lure underwater. Bottom right: a lure mimicking the visual appearance and length of a zebrafish, with an increased eye size

has a body motion would be more attractive for the zebrafish. This is the same lure that was tested in the experiment presented in Chap. 9.

- Zebrafish lure: A lure mimicking the visual appearance of the zebrafish, made of a zebrafish image printed on a paper that was covered with a latex coating. The size of the eye was also slightly increased as suggested by Landgraf et al. (2016). The external shape, however, did not precisely reproduce a zebrafish.

The three first conditions were repeated 10 times for a total of 30 trials, with an observation time of 30 min. The last two conditions were repeated four times, also with an observation time of 30 min. This is due to the fact that we preliminary observed that the biomimetic lures obtained narrower distributions compared to the non-biometric lures, and we used a power analysis to compute the sample size required to obtain a significant result. This also explains why further tests involving biomimetic lures will use also this range of repetitions. For each condition, the lure moved CW half of the time and CCW half of the time. To ensure that fish were not exposed to the same stimuli twice, we randomly selected the zebrafish in the housing aquarium each morning to perform the experiments.

10.4.2 *Results and Discussion*

A comparison of the collective decisions of the zebrafish shoal for all conditions tested in experiment 1 is shown in Fig. 10.6.

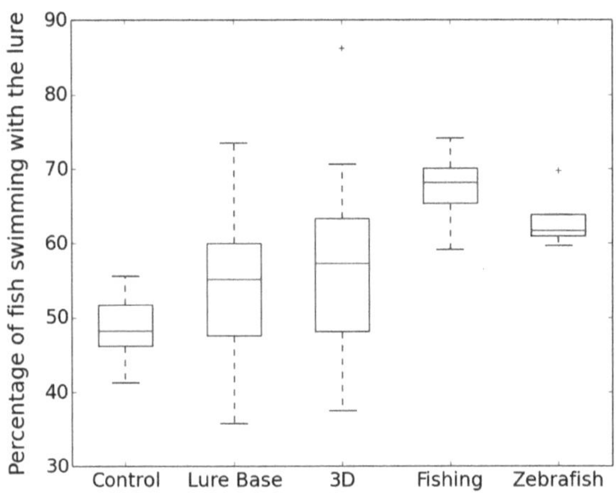

Fig. 10.6 Zebrafish swimming direction preference for experiment 1 for the five different conditions tested. Each of the first three conditions was tested ten times, and the last two conditions four times, with a trial duration of 30 min. The zebrafish were randomly selected from their housing aquarium. The zebrafish were used only for one experiment every day, but it is probable that some zebrafish were chosen to perform several conditions. The distributions differ significantly (Kruskal–Wallis $p < 0.01$), showing that the lure can have an impact on the zebrafishes' swimming direction preference

We found a significant difference for the distribution between the five focal groups (Kruskal–Wallis, $p < 0.01$), and therefore we concluded that the shape of the lure has an impact on the collective decision of the zebrafish. A post-hoc analysis demonstrated that the mean ranks of the conditions with the fishing and zebrafish lures significantly differ from the control experiment, which is not the case for the lure base and the 3D lure.

For the module and the 3D printed lure we can observe that they have a large distribution. Moreover, the zebrafish sometimes even swim in the opposite direction of the lure. For these two cases, we conclude that the lures are not very attractive, despite the fact that the average of the fish swimming direction preference is above 50% as compared to the lure.

For the fishing lure and the zebrafish lure, we have distributions that are narrower than for the module and the 3D lure, and the mean is also higher. We conclude that for these conditions, the biomimetics lures are more attractive, as the fish have a tendency to swim with them most of the time. The best result in terms of the mean was obtained for the fishing lure. Even though it does not mimic the visual appearance of the zebrafish in particular, it does attract them more. This can be explained by the fact that the tail of the lure is beating passively, compared to the zebrafish lure. This was one of the reasons the robotic lure RiBot was designed with a beating tail.

10.4.3 Tests Using idTracker

We performed a test using the software idTracker on high-definition videos in order to determine whether the effect of the lure on the fish is due to the lure itself or other factors.

Figure 10.7 shows the linear speed of the fish for experiments with and without a fishing lure moving at $12\,\mathrm{cm\,s^{-1}}$. It can be seen that, despite the fact that the lure seems to have an influence on the swimming direction of the zebrafish, the linear speed of the zebrafish is not influenced (ANOVA, $p > 0.05$).

Figure 10.8 shows the percentage of swimming direction for individual zebrafish versus the median distance between individual zebrafish and the lure for experiments involving 5 zebrafish and the fishing lure. It can be noticed that the fish that are the closest to the lure during the experiment are also the zebrafish that swim in the same direction. We performed a linear regression fitting that showed that the distance from the lure has a significant impact on the individual zebrafish swimming direction preference ($p < 0.01$). We can conclude that the attraction of the lure is increased when the fish are close to the robot. Thus, the lure seems to influence the zebrafish more than other factors.

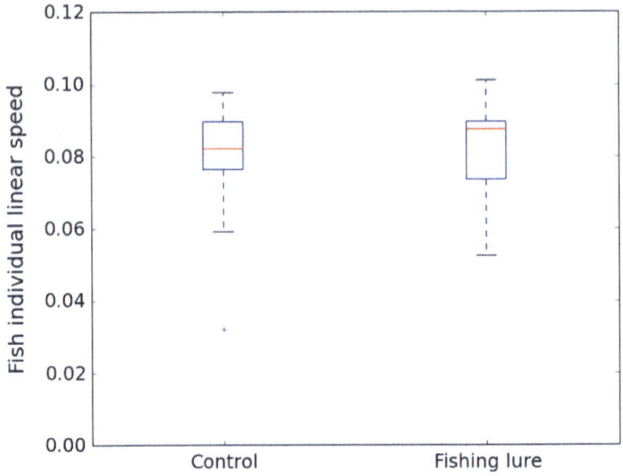

Fig. 10.7 Linear speed difference of the zebrafish between the experiments involving the lure and experiments without anything moving in the circular corridor. We tested each condition three times, with a trial duration of 30 min. The two distributions do not differ significantly (ANOVA, $p > 0.05$), which indicates that the moving lure does not influence the average linear speed of the fish

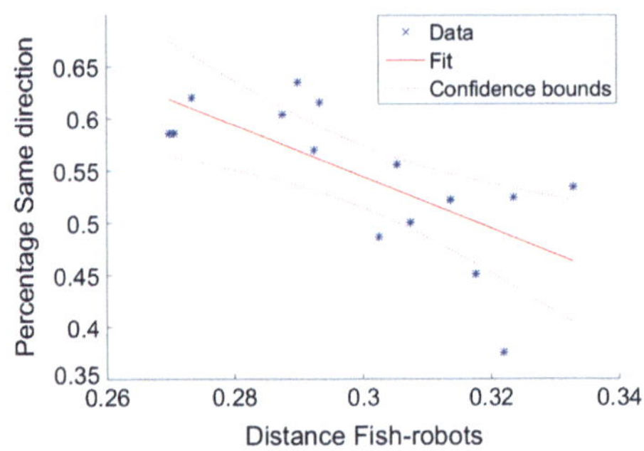

Fig. 10.8 Relation between the individual fish-robot distance and the fish swimming direction for three experiments involving five fish and the fishing lure moving in the circular corridor during 30 min. A linear regression fitting (red line) shows that there is a significant impact of the distance from the lure on the individual fish swimming direction ($p < 0.01$)

10.5 Experiments 2: Testing the Number of Lures

10.5.1 Experimental Design

In this experiment, we compare the effect of a different number of lures attached to the same module in regard to their attraction to the zebrafish shoal. Indeed, having more lures attached to the same module could increase its attraction, as the zebrafish could be more influenced by a shoal of lures than by an individual one. Also, the water

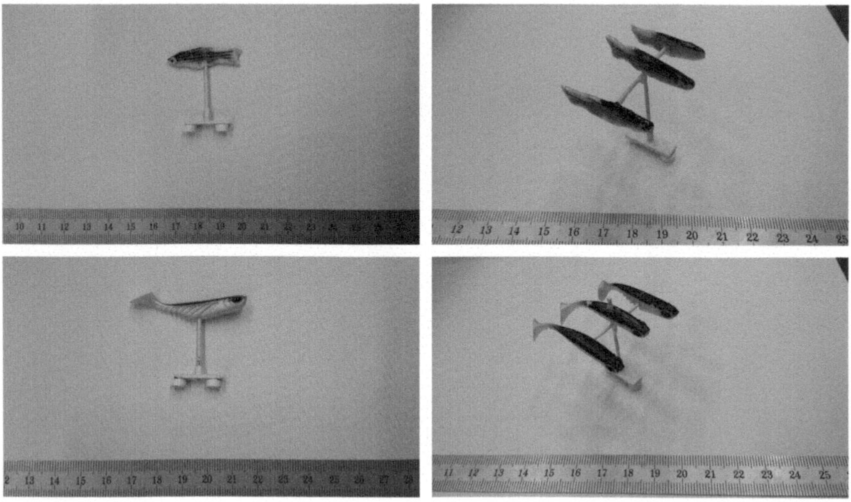

Fig. 10.9 Lures that were used in Experiment 2. Top left: lure mimicking the zebrafish size aspect ratio and visual appearance. Top right: three lures mimicking the zebrafish size aspect ratio and visual appearance. Bottom left: a fishing lure mimicking the fish size aspect ratio and beating the tail passively during the motion of the lure underwater. Bottom right: three fishing lures mimicking the fish size aspect ratio with the tail beating passively during its motion underwater that was used in Experiment 1

drag and water current generated by three lures is greater than the one generated by an individual lure, which might impact the collective preference of the zebrafish.

Four types of conditions were tested using the lures presented in Fig. 10.9:

- 1 zebrafish lure: The same zebrafish lure that was tested in Experiment 1
- 3 zebrafish lures: 3 zebrafish lures attached to a base composed of two carbon sticks, one vertical and one horizontal, glued to the vertical one
- 1 fishing lure: The same fishing lure that was tested in Experiment 1
- 3 fishing lures: 3 fishing lures attached to a base composed of two carbon sticks, one vertical and one horizontal, glued to the vertical one.

10.5.2 Results and Discussion

Figure 10.10 shows the result of experiment 2, with a comparison between the individual fishing and zebrafish lures and the modules with three of these two lures. We can observe that in terms of averages and distributions, the module with three lures does not result in a significant change in the attraction of the lure to the zebrafish shoal compared to the module equipped with an individual lure. A Kruskal–Wallis test was run on the data ($p = 0.63$) and a post-hoc analysis that confirmed that there are

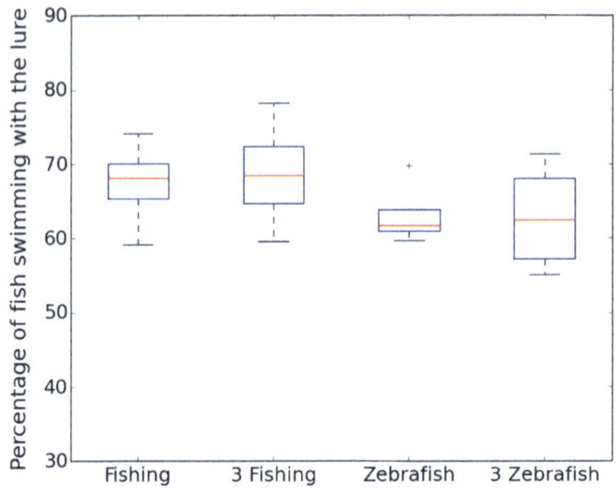

Fig. 10.10 Fish swimming direction preference for experiment 2 for the four different conditions tested. Each condition was tested four times with an experiment duration of 30 min. The fish were randomly selected from their housing aquarium. The fish used for only one experiment. The distributions do not differ significantly (Kruskal–Wallis $p > 0.05$), showing that the number of lures does not seem to have an impact on the swimming direction preference of the zebrafish

no significant differences between the four distributions. However, it seems that there is a small preference for the fishing lure compared to the zebrafish lure, which was also shown in Experiment 1, but is confirmed here using the experiments involving multiple lures.

The results obtained in this experiment will be used for the design of further experiments with mixed societies of fish and robots. An increase in the number of lures on a module moving at the same speed, such as the lures used in Ladu et al. (2015b), Bartolini et al. (2016), does not seem to have an impact on the attractiveness of the module. Tests should be performed with lures moving at different speeds and trajectories to examine the impact on the fish collective decision. Indeed, in this experiment, the three lures were moving together closely, forming a small swarm of fish. Fish are probably more attracted by lures moving like them, i.e., in a shoal formation. The fact that three lures are generating more water current and water drag does not seem to have a significant impact on the collective decision of the zebrafish to follow the lures.

10.6 Experiments 3: Lures Moving in Opposite Directions

10.6.1 Description of Experiments

As we have shown in the description of the experimental setup, the system designed to rotate the lures is equipped with two rotors, and thus it is possible to rotate two lures in the same or in the opposite directions. Here, experiments were performed with two lures moving in opposite direction, one close to the inner boarder of the corridor and one near the outer boarder of the corridor. This test was done to determine if one lure is more attractive than the other in the same experiment.

Two types of conditions were tested:

- 1 zebrafish lure versus 1 lure base
- 1 zebrafish lure versus 1 fishing lure.

10.6.2 Results and Discussion

The results for experiment 3 are shown in Fig. 10.11. We ran a Kruskal–Wallis test to prove that there was a significant difference between the two distributions ($p < 0.05$). Thus, we can conclude that, compared to a zebrafish lure, the lure base and the fishing lure have a significantly different effect on the collective decision of the zebrafish.

This test could thus confirm the results obtained in experiment 1. Indeed, the fish seem to have a preference for the fishing lure compared to the zebrafish lure, and they have a clear preference for a biomimetic lure compared to the lure base.

This methodology is therefore appropriate for finding the most attractive lure among different lures. Moreover, here, the water current generated by one of the lures moving is also partially cancelled by the water current generated by the lure moving in the other direction.

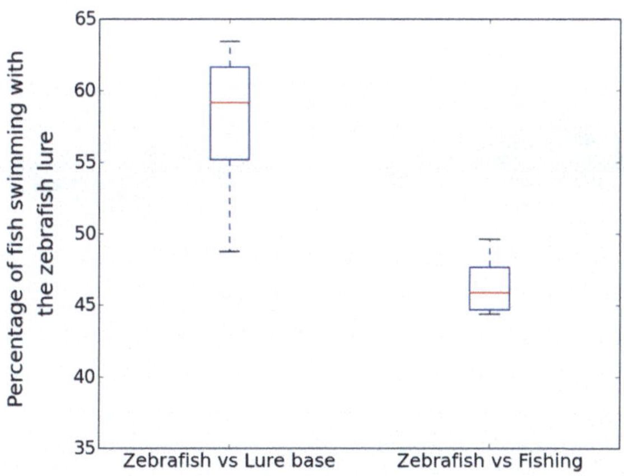

Fig. 10.11 Results for experiment 3. For the first condition, a zebrafish lure was moving in one direction and the lure base in the opposite direction. For the second condition, the zebrafish lure was moving in one direction and the fishing lure was moving in the opposite direction. Each condition was tested four times. The fish were randomly selected from their housing aquarium. The fish used for only one experiment. The distributions significantly differ (Kruskal–Wallis $p < 0.05$), showing that the zebrafish have different preferences for different lures moving in opposite directions

10.7 Experiments 4: The Effect of a Biomimetic Actuated Lure

10.7.1 Experiments Description

Based partially on the results obtained in experiments 1, 2, and 3, and on the current state of the art in fish-robot interaction, we designed the robotic lure RiBot mimicking the visual appearance of the zebrafish (Chap. 5). Additionally, this lure is able to beat its caudal peduncle with a large range of amplitudes and frequencies.

In this experiment, we tested the effect of the actuated tail and the color appearance of the RiBot lure on the collective decision of the zebrafish. The tail was actuated with a frequency of 5 Hz and an amplitude of $\pm 5^{\circ}$. The choice of the tail beating frequency and amplitude was made based on preliminary experiments that were presented in Bonnet et al. (2016b) that tested the attraction of the lure towards fish for different ranges of frequencies and amplitudes. The two RiBots that were used during this experiment can be seen in Fig. 10.12.

The RiBot can beat its tail thanks to a stepper motor, which also generates some noise when activated. First, we tested if the noise of the motor has an effect on the collective decision of the zebrafish. We decoupled the caudal peduncle part from the motor in order to activate the motor without generating any movements of the tail. We tested two conditions:

- lure without pattern and motor not moving
- lure without pattern and motor moving without the tail fixed to the motor

Fig. 10.12 Actuated biomimetic lures used in experiment 3. Left: lure with a zebrafish color pattern. The pattern was printed on the skin of the lure using decals. The pattern was then covered with latex to become waterproof. Right: lure without any zebrafish color pattern. The PCB with all electronic component that are used for the actuation of the caudal peduncle can be seen from outside

Then, we tested four conditions in order to determine the attractive factors of the RiBot among the visual appearance and the tail beating:

- lure with pattern and tail not moving
- lure with pattern and tail moving
- lure without pattern and tail not moving
- lure without pattern and tail moving.

10.7.2 Results

Figure 10.13 shows a comparison between experiments in which the motor was not activated and experiments with the motor activated without the tail beating movements. It shows that the motor-generated vibrations have no significant effect on the swimming direction of the zebrafish (Kruskal–wallis $p > 0.46$), and thus we conclude that the noise of the motor has no effect on the behavior of the zebrafish.

Figure 10.14 shows the result of the test with the four conditions: RiBot with pattern and tail on, RiBot without pattern and tail on, RiBot with pattern and tail off, and RiBot without pattern and tail off. It appears that both the pattern and beating of the tail seem to have an effect.

We computed the relative effect of each factor: the effect of the motor a_1, the pattern a_2, and the combination of both effect a_{12}. Table 10.2 shows the relative effects of the factors. One can observe that both a_1 and a_{12} have a relative effect

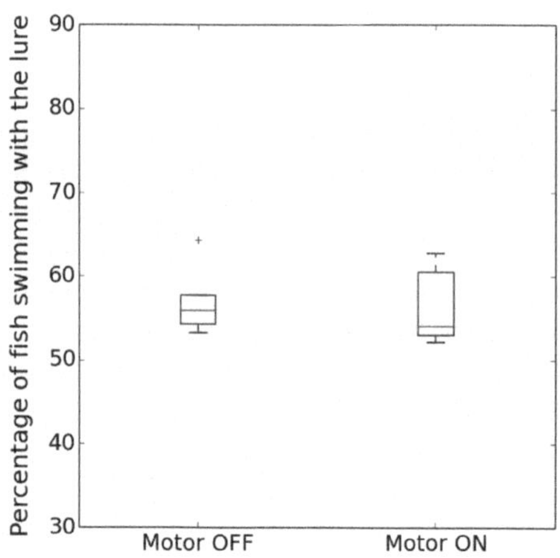

Fig. 10.13 Effect of the noise of the motor on the swimming preference of the zebrafish. Each condition was tested six times with a trial duration of 30 min. The fish were randomly selected from their housing aquarium. The zebrafish were used for only one experiment. The distributions do not significantly differ (Kruskal–Wallis $p > 0.05$), showing that the noise made by the motor does not significantly change the swimming direction preference of the fish

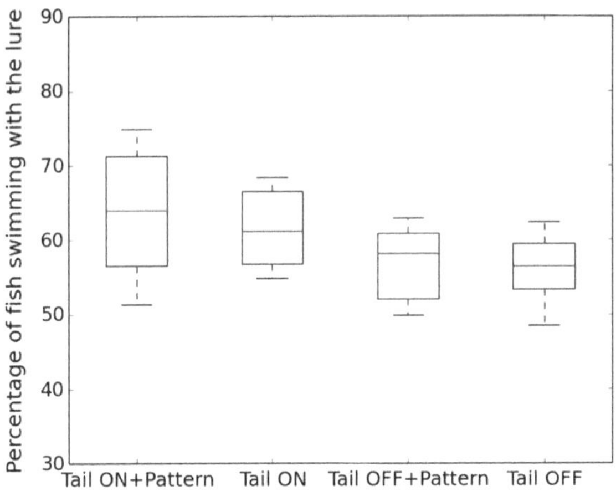

Fig. 10.14 Fish swimming direction preference for experiment 4 with the four conditions tested: the lure with a pattern beating the tail (Tail ON+Pattern), the lure without pattern beating the tail (Tail ON), the lure with a pattern not beating the tail (Tail OFF+Pattern) and the lure without pattern not beating the tail (Tail OFF). Each condition was tested six times. The fish were randomly selected from their housing aquarium. The fish were used for only one experiment per day, but it is probable that the same fish was used to test several conditions. The analysis of the relative effect shows that the beating tail seems to have the most effect on the attractiveness of the RiBot to zebrafish

Table 10.2 Relative effect of the different factors on the swimming direction of the zebrafish

Factor	ID	Relative effect (%)
Tail beating	a_1	4.54
Pattern	a_2	0.74
Tail beating and pattern	a_{12}	2.97

close to 5%, compared to a_2 which seems to have a lower effect on the swimming direction of the zebrafish. This is consistent with the results obtained in experiments 1, 2, and 3, which showed that the fishing lure that is equipped with a beating tail is more attractive than a lure mimicking only the appearance of the zebrafish. However, the visual appearance also plays a role, as shown in experiment 1: A lure simply mimicking the shape of a zebrafish has no significant attraction, and thus both factors, the visual appearance and the beating tail, should be probably considered in the design of a robot that has to interact with fish. We performed an ANOVA on the data, the results of which can be seen in Table 10.3. The only factor that has a significant impact is the beating tail a_1 ($p < 0.05$), confirming that this factor plays a non-negligible role in the attraction of the zebrafish.

10.7.3 Tests Using idTracker

Figures 10.15 and 10.16 show the percentage of swimming direction for individual fish versus the median distance between individual fish and the RiBot for experiments

Table 10.3 ANOVA obtained for experiment 4. a_1 corresponds to the beating tail factor, a_2 to the color pattern factor, and a_{12} to the interaction term

Effect	SS	df	MS	F	p
a_1	265.07	1	265.069	6.79	0.0169
a_2	4.05	1	4.051	0.1	0.7506
a_{12}	16.1	1	16.105	0.41	0.5279
Residual	780.29	20	39.015		
Total	1065.52	23			

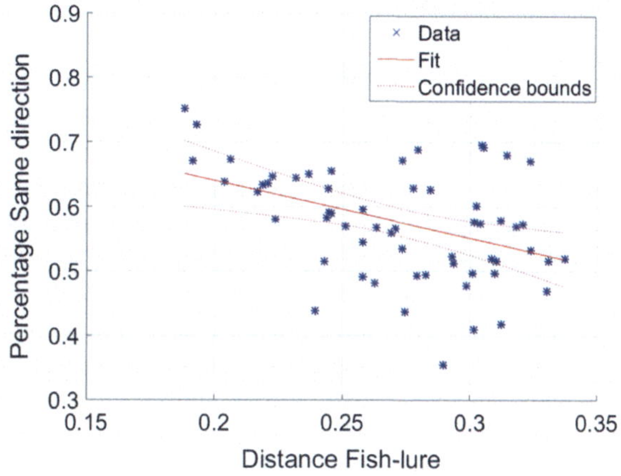

Fig. 10.15 Individual fish-RiBot distance versus the fish swimming direction for RiBot ON for the experiments involving five zebrafish and the actuated lure moving in the circular corridor with the tail beating. The red line indicates the linear regression that was performed on the data

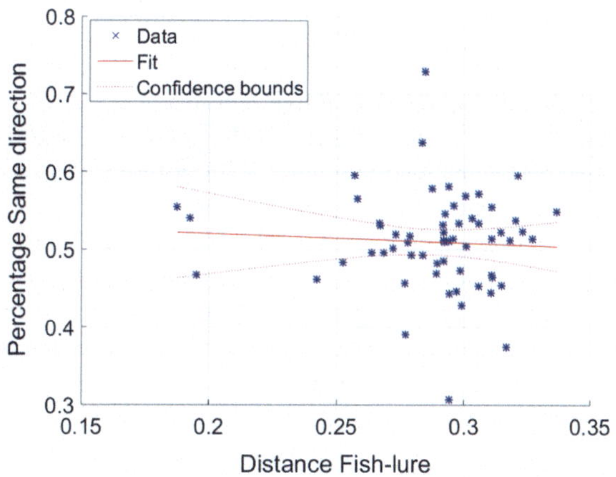

Fig. 10.16 Individual fish-RiBot distance versus the fish swimming direction for RiBot OFF for the experiments involving five zebrafish and the actuated lure moving in the circular corridor with the tail not beating. The red line indicates the linear regression that was performed on the data

involving five zebrafish and the RiBot that has the tail ON (Fig. 10.15) and the tail OFF (Fig. 10.16). We can observe that the experiment with the lure beating its tail offers a result more similar to what was shown in Fig. 10.8, with a correlation between the individual fish-lure distance and the fish swimming direction. However, this seems not to be the case for experiments with the RiBot not moving its tail. A linear fitting regression was applied to the two different conditions, showing that there is a significant correlation between the fish-lure distance and the fish swimming direction in experiments with the tail of the RiBot ON ($p < 0.05$), which is not the case for experiments with the RiBot with its tail OFF ($p = 0.668$). Thus, for an attractive lure, we can confirm that the fish that are close to the lure are also the ones that are mostly swimming with it.

10.8 Conclusion

In terms of the method that was presented in this chapter, we show that the system developed with the coaxial motor does not affect the swimming direction of the zebrafish if the motor is rotating without being coupled with any object inside the tank. We also showed that, inside a circular corridor, without any object rotating, and during experiments of 30 min, a shoal composed of five zebrafish will have a tendency to swim half of the time CW and half of the time CCW.

By introducing a constant water current inside the corridor, we could significantly change the swimming direction preference of the zebrafish shoal, with the fish shoals that preferred the direction always opposed to the water current in all the experiments. This indicates that a water current induced by an object moving inside the corridor can change the collective behavior of the zebrafish. However, if the moving object was attractive to the zebrafish, there should be a counter-effect, with the fish starting the swim in the direction of the current generated by the lure.

When having a lure inside a corridor moving at a speed above the average linear speed of the zebrafish, we could trigger an effect on the zebrafish shoal's direction of swimming. A shoal size of five zebrafish was a good tradeoff between homogeneity and the effect on the collective behavior that we wanted to observe.

Experiments involving lures with different shapes confirmed that the lures with the most biomimetic cues are the most attractive for the zebrafish shoals. This result is not new when compared to what can be found in literature. However in this study, we wanted to decouple all the influencing stimuli that could be perceived by the fish. We wanted to be sure that this effect was not due to the shape of the lure base or to a generated water current. We could indeed see that a small white module moving at high speed could also change the swimming direction of the fish. However, the distribution was very large, indicating that in some cases it can influence them, but the effect is not as significant as it is for the lures with a biomimetic appearance.

The use of the software idTracker confirmed that the methods used in the online tracking to determine the swimming direction of the zebrafish was accurate. Videos processed with idTracker also showed that the linear speed of the zebrafish did not change between experiments involving the lure and experiments without any objects moving underwater. This indicates that the lure was not introducing more stress for the fish nor changing their behavior drastically. Finally, we could show that there was a correlation between the individual fish-lure distance and the fish swimming direction when the fish where swimming more in the direction of the lure. This shows that the closer the fish are to the lure the more they are swimming in the same direction. Hence, the decision of the zebrafish is not influenced by the fact that they are frightened by the lure; it is the lure itself that influences the fishes' decision.

We also show that increasing the number of lures on the support does not influence the decision of the shoal, despite the fact that the lures were moving in the same direction and generated more water perturbations than an individual lure.

Finally, we studied the effect of an actuated biomimetic lure, using the beating of the tail and the coloration as parameters to study the attractiveness of the device. We show that the tail beating has a significant effect on the attraction of the fish towards the lure, compared to the coloration.

All these results confirm that a lure designed to perform experiments involving mixed societies of fish and robots should include two main features in order to increase the attraction of the lure: a biomimetic shape and a beating tail.

Chapter 11
Towards Mixed Societies of Fish and Robots

11.1 Summary

In the first chapter of this dissertation, we defined the main goal of this study: successfully designing robotic tools to create mixed societies of robots and fish. Then, we showed how we designed the different robotic devices and controllers, using experiments involving fish to assess and improve their efficiency to trigger the social behavior of zebrafish.

The aim of this chapter is to show how we can use the developed robotic system in experiments involving mixed societies of fish and robots (Fig. 11.1). These experiments were conducted to show that a group of robotic devices could be integrated into zebrafish shoals and are also able to monitor the shoals collective decisions.

11.2 Methods

11.2.1 Binary Choice Setup

This chapter is based on a methodology similar to the one used in the LEURRE project, where the authors succeeded in implementing a mixed society of living and artificial agents (cockroaches and robots, respectively) (Halloy et al. 2007). The goal behind this methodology is twofold: first, to show that the artificial agents behave like the group of animals, and second, that the mixed society can behave in the same way as if the society was composed of only real animals. Once we have shown that the artificial agents are integrated into the society, we must show that they can also influence the society in its collective decisions, as it is the case for some members of the animal society that can make decisions for the collective (Krause et al. 2000; Ward et al. 2013).

F. Bonnet and F. Mondada, *Shoaling with Fish: Using Miniature Robotic Agents to Close the Interaction Loop with Groups of Zebrafish Danio rerio*, Springer Tracts in Advanced Robotics 131, https://doi.org/10.1007/978-3-030-16781-3_11

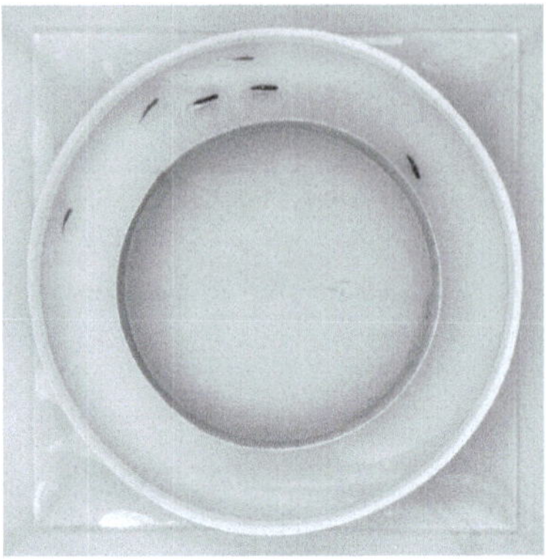

Fig. 11.1 Top view of the setup that was used to demonstrate the creation of a mixed society composed of three robots and three zebrafish

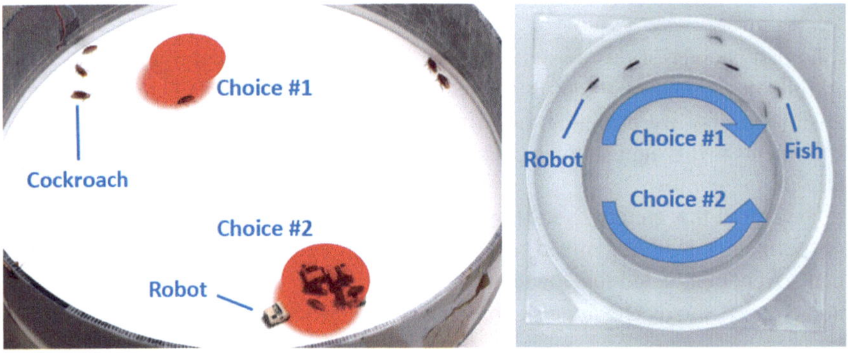

Fig. 11.2 Binary choices for LEURRE project and the mixed society of fish and robots. Left: Experimental setup used in the LEURRE project, with a binary choice made of two shelters. The robots and the cockroaches chose one shelter under which to aggregate after a certain period of time. Right: The setup that we used for the experiments presented in this chapter, where the binary choice is the swimming direction of the mixed society of fish and robots. Here the fish and robots agreed on the swimming direction (CW)

In order to show these two experiments, a binary choice setup was specifically designed for the LEURRE project (Fig. 11.2, Left). In the first step, the setup was designed to offer an equal choice between two spots for the animals to test whether the robots were well integrated into the animal society. The setup could also be biased in order to observe how the robot could force the animals' choices. In the LEURRE project, the binary choice was composed of two shelters placed in an open arena under which the cockroaches could hide. The setup could be biased by changing the darkness of one of the shelters, as the cockroaches would then choose to aggregate more often under the darker shelter.

11.2.2 Arena Selected to Conduce Mixed Society Experiments

As we have already shown in Chap. 10, it is very difficult to design a setup where the location of the zebrafish can be stabilized for a long period of time, as they are constantly moving and exploring the arena. Therefore, in order to design an experiment with a mixed society of robots and fish in which the collective choices made by the fish shoals can clearly be measured, we have used the same arena as the one presented in Chap. 10, with some adaptations (Fig. 11.2, Right). We showed in Chap. 10 that the circular corridor setup offered an equal binary choice, as the fish shoals will tend to swim clockwise (CW) half of the time and counter-clockwise (CCW) half of the time. Thus, this setup is suitable for testing whether the robots can be integrated into the fish society without changing their collective behaviors and if the robots can influence these behaviors, using the results that were obtained in Chap. 10. Additionally, we needed a biased setup to test whether the robots could force the zebrafish to move in a certain direction that would not be preferable for them.

In order to bias the experimental setup, we designed an apparatus placed in the middle of the circular corridor arena that can rotate a pattern made of black stripes on a white background (Fig. 11.3). The visual behavior of zebrafish has already been deeply studied (Fleisch and Neuhauss 2006), and some studies have shown that they can react differently to black and white stripes moving at different speeds (Maaswinkel and Li 2003). We used this mechanism as an environmental parameter that could bias the collective choice of the fish without using the robotic agents.

The designed system consists of a cylinder with an outer diameter of 38 and 12 cm in height. Inside the cylinder, a DC motor with a gearbox (Faulhaber, Switzerland) is rotating a rotor on which the pattern made of black and white stripes is attached. A raspberry PI is used to control the rotation speed of the motor in a closed-loop, using the optical encoder of the motor as a measurement for the speed. A Pulse Width Modulation (PWM) signal at 1 KHz is generated on a Raspberry PI GPIO to control the motors via a SI9989 H-bridge motor driver. The speed is controlled with a Proportional-Derivative (PD) controller. The system is powered by an external USB rechargeable battery. The autonomy of the system is about three hours, and the device can reach speeds between 0 and 30 revolutions per minute (rpm). The motorization is very silent and can be barely heard by a human ear. The system can be remotely controlled via Bluetooth, and the operator can select the speed and the direction of rotation during the experiment.

We performed preliminary tests to determine the rotational speed and the width of stripes that could affect the collective behavior of the fish. Among the different values that were tested, we selected a speed of 12 rpm and a width of the black and white stripes of 3 cm, which resulted in a clear bias in the swimming direction of the zebrafish, as it will be shown in Sect. 11.3.1.

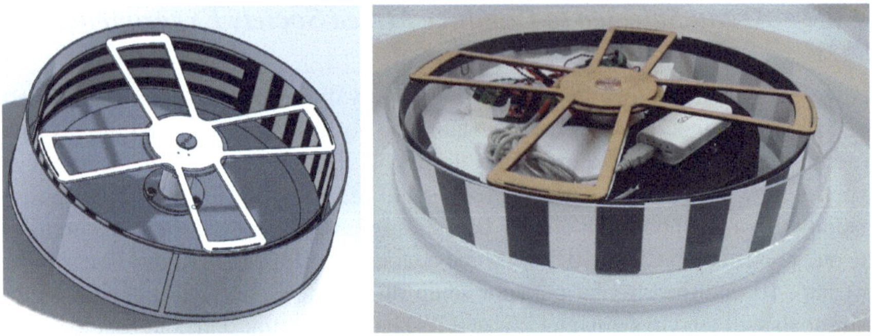

Fig. 11.3 Left: 3D CAD design of the rotating stripes machine. A coaxial motor is rotating a rotor on which a circular sheet composed of black and white stripes is attached. Right: The rotative stripes machine installed in the circular arena. The system is powered by a rechargeable LiPo battery and controlled by a Raspberry PI. The Raspberry PI can be remotely controlled from a computer via Bluetooth

11.2.3 Mixed Society Size

In Chap. 10, we showed that the size of the fish shoal should be chosen as a function of the group homogeneity and the attraction effect of the lures on the fish group. As we performed most of our experiments in Chap. 10 with groups of six agents (five fish responding to one lure), we decided to use a mixed society composed of six agents, three fish and three robots, and compared it with a control experiment where the group was composed of six zebrafish.

11.2.4 FishBot Control

Compared to Chap. 10, in which a coaxial motor was rotating the lure inside the circular corridor, here, in order to build a mixed society composed of fish and multiple robots, we used the FishBots to move the lures. Due to their small width, two FishBots can cross inside the circular corridor, and thus the robots can achieve the same types of movements as the fish do in this setup.

The three FishBots were controlled in a closed-loop thanks to CATS, using the PID control described in Chap. 7 with a linear speed of $8\,\mathrm{cms}^{-1}$, which forced the FishBots to rotate either in a CW or CCW direction. The Braitenberg local obstacle avoidance allowed the FishBots to avoid the walls. We observed that the resulting behavior mimicked the behavior of zebrafish swimming in a narrow corridor, with the robots swimming along the corridor walls with small oscillation between the two walls.

11.2.5 Lures

For this experiment, we used the fishing lure that we presented in Chap. 10 that demonstrated the best acceptance by the fish society. The lures that were coupled with the FishBots were all identical.

11.2.6 Experiment Design

In both the biased and non-biased setups, we tested three conditions:

- **Control (CT)**: Six zebrafish were placed in the circular arena without any robots and were free to move either CW or CCW. The zebrafish were thus only influenced by the environment, i.e., any bias in the swimming direction could be contributed to the rotative stripes machine.
- **Robots swimming with fish (RW)**: In this experiment, groups of three robots and three zebrafish were tested. The robots were following the decision made by the shoal of fish, which was determined by CATS and the same algorithm that was used to determine the fish shoal swimming direction presented in Chap. 10. The robots were controlled to always swim in the same direction; thus, if the majority of fish decided to swim in one direction, the three robots were controlled to swim in the same direction.
- **Robots imposing a choice (RI)**: In this experiment, three robots were placed with three zebrafish, and the robots were controlled to turn only in one direction in the circular corridor. Hence, it was an extreme case that measured the impact of three robots going in one direction on a shoal of three fish. The rotating direction of the FishBots was varied between the experiments in order to eliminate any bias. Even though the robots were programmed to move in only one direction, in some cases they were not turning in the same direction 100% of the time, as they sometimes collided with each other, which could cause one or two robots to move in the wrong direction for a short amount of time. However, as we obtained a common direction 95% of the time, we assumed that they were mostly moving in one direction.

The experiments lasted for 40 min. In the first 10 min, we let the fish adapt to the new environment, and then we started the experiment, which we recorded with the Basler camera in high definition for post processing using idTracker (as described in Chap. 10). We repeated each condition eight times. We had 60 zebrafish at the time of the experiment. The zebrafish were randomly selected from their housing aquarium in the morning to perform a set of five to seven trials. We also varied the bias, i.e., the swimming direction of the robots for the third condition, as well as the direction of rotation of the rotative stripe device for randomization in order to avoid other biases.

11.2.7 Measurements

We used the same algorithm for the measurements that was used in Chap. 10 to determine the swimming direction of the shoals, which was also used as input for the closed-loop control of the robot. We also used idTracker to determine the individual trajectories and speeds. In these experiments, we also made a measurement of the two species separately (robots and fish) as well as of the mixed society (fish with robots). For instance, we will describe in further sections the second condition RW by mentioning the whole group of six agents (RW6A), only the three fish without taking the robots into account (RW3F), and only the three robots without taking the fish into account (RW3R).

11.3 Results and Discussion

11.3.1 Collective Decision

Non-biased Setup

Figure 11.4a, shows the percentage of swimming direction for the entire shoal (six agents) in the non-biased setup. In the first condition, the control (CT), the group consists of six zebrafish, and in the two other conditions (RW6 and RI6), it consists of mixed societies of three fish and three robots. We can observe that the decision of the mixed society of three fish and three robots is the same for condition RW as for the condition CT. This was predicted, as the robots should be integrated and not change the collective decisions of the zebrafish. On the other hand, we can observe for the third condition (RI6) that the three robots, by swimming in only one direction, could modify the swimming direction preference of the whole mixed society compared to the RW6 condition. The three distributions are significantly different (Kruskal-Wallis $p < 0.05$), and a post-hoc analysis shows that the mean ranks of distribution of conditions CT and RW6 are significantly different from that of condition RI6, while condition CT and RW6 have no significantly different distribution.

If we look at the two species separately for conditions RW and RI (Fig. 11.4c, e), we can observe that for condition RW, the zebrafish seem to behave in the same way as in condition CT, swimming half of the time in each direction, which influences the robots to behave as programmed and to follow the estimated collective decision of the zebrafish. This result shows that the zebrafish were not perturbed in their collective decision in this condition. Figure 11.5, left, shows the percentage of swimming direction for each trial separately for the RW condition. It shows that the robots were able to follow the decisions of the zebrafish in the non-biased setup most of the time.

If we now observe the separation of both species for condition RI (Fig. 11.4e), we can see that the swimming direction preference of the zebrafish was changed, and they were mostly swimming in the same direction as the robots. This result is

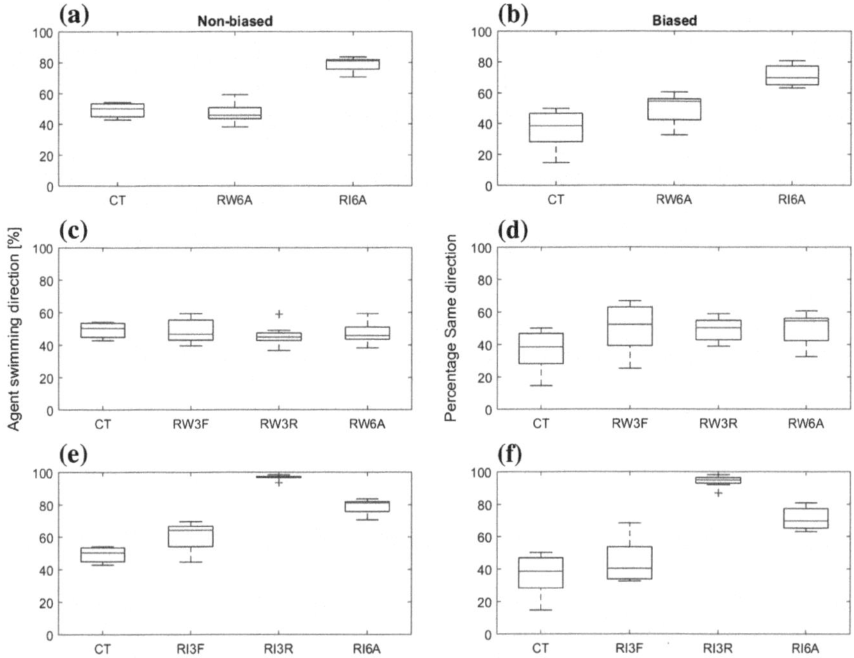

Fig. 11.4 Mixed society swimming direction preference for the three tested conditions in a non-biased (Left) and biased setup (Right). **CT** stands for control experiment condition, **RW** robots swimming with fish, and **RI** robots imposing a choice. We also make the distinction between the group of six agents (6A) and the subgroups of three fish (3F) and three robots (3R) for the second and third conditions. Eight trials were performed for each condition, and the duration of each trial was 30 min

also very promising as it shows that the robots are able to influence the collective behavior of the fish shoal in a binary choice setup. Compared to the experiments that were shown in Chap. 10, the robots here were not programmed to move faster than the average linear speed of the zebrafish. Thus, the fish were not only attracted because the lures were moving faster but also because they were all moving in one direction at the same speed as the zebrafish.

Biased Setup

Figure 11.4b shows the percentage of swimming direction for the entire shoal (six agents) for the three conditions in the biased setup. We can observe that the setup indeed induced a bias in the collective choice of the zebrafish, with the society of six zebrafish (CT) swimming with an average percentage of 60% in the opposite direction of the rotating stripes. The bias also has a strong influence on the distribution of this decision when compared with the distribution of the fish collective choice shown in Fig. 11.4a. Indeed, the zebrafish seemed to be sometimes highly affected by the rotative stripes while other times almost unaffected. We assume that the reason the fish primarily swim in the opposite direction to the rotating pattern is because the

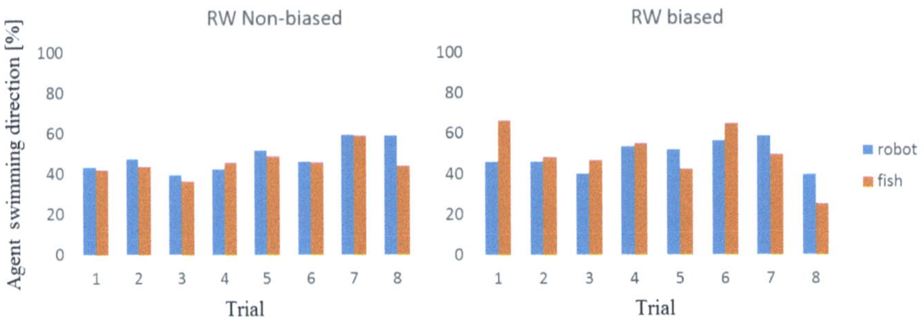

Fig. 11.5 Mixed society swimming direction preference for each of the eight trials in the non-biased (left) and biased (right) experimental setups for the second condition RW, for which the three robots were programmed to follow the collective decision of the zebrafish. In orange, the collective preference of the three zebrafish, and in blue, the collective preference of the three robots

rotative stripes simulate a water current and this effect is less perceived than a physical water flow. As shown in Chap. 10, Sect. 10.3, the fish have a tendency to swim against a water current.

When comparing the mixed society of three robots and three fish for conditions RW and RI with condition CT in Fig. 11.4b, it can be seen that the robots have an effect on the group decision (Kruskal-Wallis $p < 0.01$). Using a post-hoc analysis, we can state that condition CT and RW do not differ significantly, and they have significantly different distributions compared with condition RI. Therefore, we can assume that we can build a mixed society of three fish and three robots that, on one hand, could behave with collective decisions like a shoal of six fish in a biased setup, and on the other hand, that the robots could modify the whole mixed society by influencing the decision.

For condition RW, for which the robots were programmed to follow the collective decision of the fish, we can observe that the collective choice of the fish was modified (Fig. 11.4d). The robots in this condition seem to have stabilized the decision of the zebrafish, around 50 % in both directions, and the fish seem to be more likely to ignore the effect of the induced bias. When looking at the trials separately for condition RW in the biased setup (Fig. 11.5, Right) we see that the robots had more difficulty following the zebrafish direction compared to the experiment in the non-biased setup. This can be explained by the fact that, in the biased setup, the zebrafish move with more abrupt movements that are difficult to mimic for the closed-loop control of the robots.

For the third condition, in which the robots imposed a direction on the fish (RI), we can see in Fig. 11.4f that the decision of the zebrafish was also modified compared with the first condition CT. However, the influence of the robots was not enough to completely change the swimming direction preference of the shoal of fish compared to the equivalent experiment performed in the non-biased setup. Thus, it seems that the robots can modulate the decisions of the zebrafish in such an environment only when the robots are programmed to move like them. Indeed, for the third condition,

RI, the robots are ignoring the decision of the fish and move on their own. The result is that the zebrafish seem to be less attracted to the robots' decision compared to the second condition, RW.

The most interesting aspect of these experiments in terms of fish behavior is that we can clearly identify a change in the fish behavior when inducing the bias using the rotating stripes machine. The resulting behavior of the mixed society was also modified, and thus the system can be used to monitor the behavior of fish in different environments. We could imagine that, using evolutionary algorithms and adaptive controllers, the robots could manage to adapt to the changes in the fish behavior and possibly find out by themselves what factors should be adapted to impose a choice, depending on the environmental conditions.

11.3.2 Collective Decision Over Time

In order to measure the potential of our system for long-duration experiments, we analyzed how the collective choices of the zebrafish shoal varied during the experiments. For that, we sampled the experiments in periods of five minutes and computed the average swimming direction percentage of the three fish for conditions RW and RI. Figure 11.6 shows the results of these measurements. We can see that for the two conditions, there are no significant differences between the periods (Kruskal-Wallis $p > 0.05$ for both conditions). This means that the effect of the robots on the collective behavior of the zebrafish seems constant for experiments of 30 min.

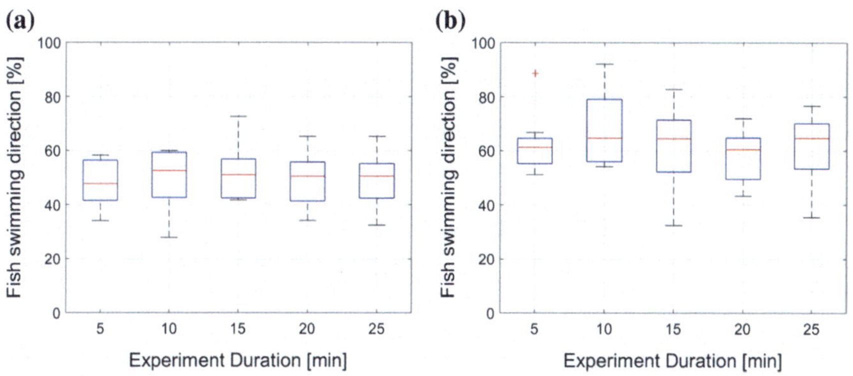

Fig. 11.6 Mixed society swimming direction preference over time for two conditions involving mixed societies of robots and fish: the robot swimming with fish (**RW**) and Robot imposing a choice (**RI**) (**a** and **b**, respectively) in the non-biased setup. For both conditions, there are no statistical differences between the periods (Kruskal-Wallis $p > 0.05$). We measured the average speed based on the eight trials that were completed for each condition in the non-biased setup

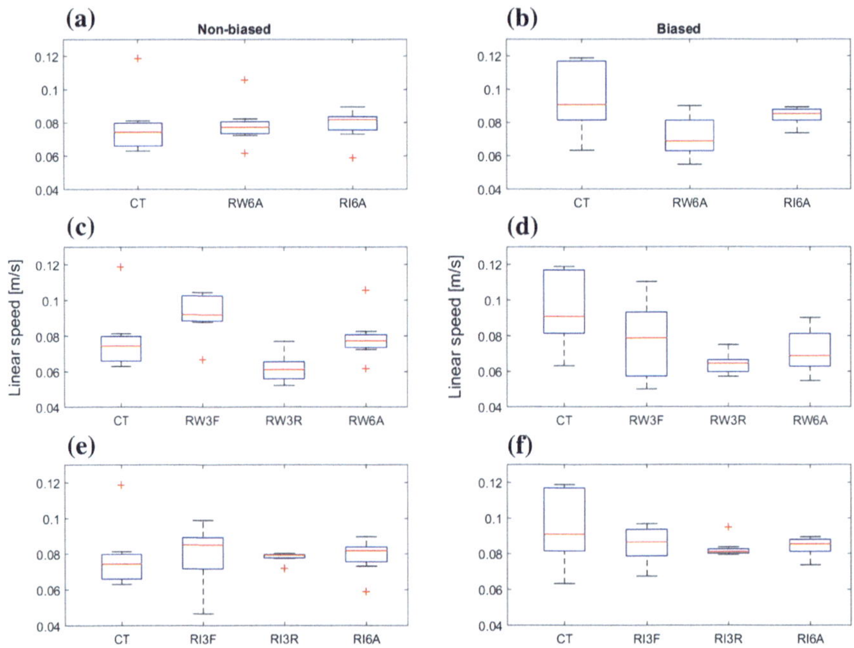

Fig. 11.7 Mean speed of the mixed society for the three tested conditions in a non-biased (Left) and biased setup (Right). **CT** stands for control experiment condition, **RW** Robots swimming with fish, and **RI** Robots imposing a choice. We also make the distinction between the group of six agents (6A) and the subgroups of three fish (3F) and three robots (3R) for the second and third conditions. Eight trials were performed for each condition, and the duration of each trial was 30 min

11.3.3 Linear Speed

We investigated other metrics in order to evaluate the impact of robots on the zebrafish behavior. We analyzed the linear speed of all the agents in the non-biased and biased setups (Fig. 11.7).

For the non-biased setup, (Fig. 11.7a), the distributions of the mean linear speed of the shoal of six agents do not significantly differ between the three conditions (Kruskal-Wallis $p > 0.05$), which indicates that the global speed of the societies in the three conditions is the same.

However, when looking at the two species of agents separately (Fig. 11.7c, e), we can observe that for the second condition, RW, the robots were moving with a mean speed smaller than that of the fish. This can be explained by the fact that the robot often turned in the setup due to the frequent change in fish swimming direction. The speed of the fish for RW3F is higher than that for the CT. This effect might be due to the fact that, by introducing robots in the group that are usually slower than the fish, the fish tend to move faster in order to maintain the same dynamics in the group. For the third condition (Fig. 11.7e), however, we can observe that the robots are moving

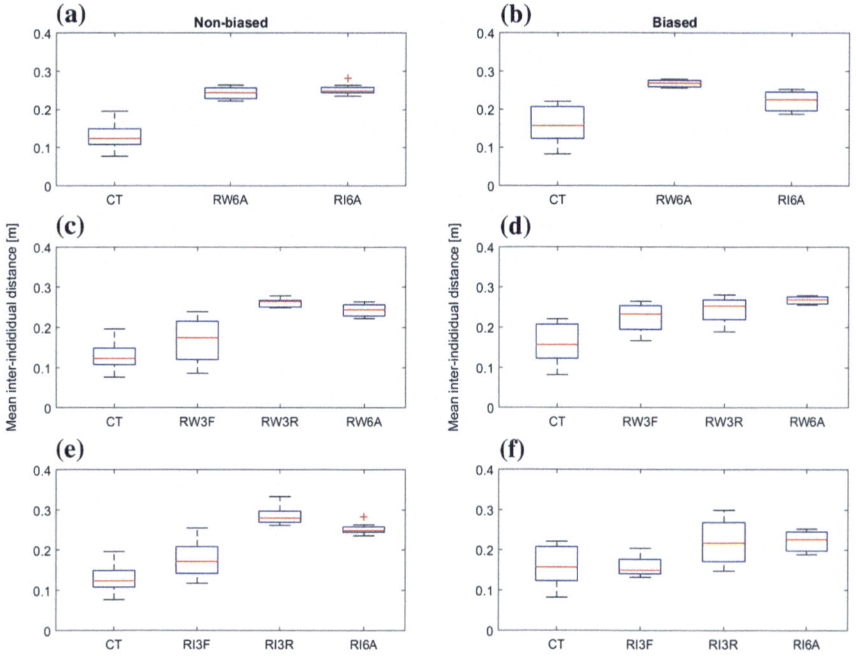

Fig. 11.8 Mean Inter-individual distances of the mixed society for the three tested conditions in a non-biased (Left) and biased setup (Right). **CT** stands for control experiment condition, **RW** Robots swimming with fish and **RI** Robots imposing a choice. We also make the distinction between the group of six agents (6A), and the subgroup of three fish (3F) and three robots (3R) for the second and third condition. Eight trials were performed for each conditions, and the duration of each trial was 30 min

at 8 cm s^{-1} on average as they are programmed to do so, and the speed of the fish is almost the same as that of the controls, CT. This result is comparable to what was obtained in Chap. 10, when we showed that a moving lure did not change the linear speed of the fish significantly.

The speed of the fish is increased in the biased setup compared to the non-biased setup, as can be seen in Fig. 11.7b. This can be explained by the fact that the fish, as they perceive that they are moving in water with a current and thus increase their linear speed to compensate for the current. The presence of the robots inside the mixed society seems to stabilize the speed of the fish (Fig. 11.7d, f).

11.3.4 Inter-individual Distance

We also analyzed how the inter-individual distances between the fish varied in the different experimental conditions. We measured three types of inter-individual distances for the three conditions:

1. Inter-individual distance between all the agents for CT, RW, and RI
2. Inter-individual distance between the three fish only for RW3F and RI3F
3. Inter-individual distance between the robots only for RW3R and RI3R

In order to compute the inter-individual distance, we computed at each time-step, for each agent, the average of the distance that separates it from the other two or five agents, and then computed the mean for all agents at the end of the experiment.

Figure 11.8 shows the mean inter-individual distance for the experiment in the non-biased and biased setups. First, we can see that the inter-individual distance of the robots is larger than that of the fish. This is due to the fact that we do not control the inter-individual distances between the robots. This could be done for further analyses, by adding controllers that make the robots swim in shoals in this experiment, using for instance the vision-based model described in Chap. 8.

Regarding the fish inter-individual distance, we can see that it does not vary greatly between experiments involving robots and experiments involving only fish. The presence of the robots moving inside the groups probably affects the cohesion of the zebrafish slightly, but not significantly. Thus, we can assume that the robots did not introduce a drastic change in the social behavior of the fish.

11.4 Conclusion

The control experiment shows that, as shown in Chap. 10, the zebrafish are moving CW half of the time and CCW half of the time in the non-biased circular arena. When a device rotated black and white stripes in one direction, the zebrafish tended to swim in the opposite direction. This result is not surprising, compared to what was shown in Chap. 10, Sect. 10.3, when a current was generated in the circular corridor, the fish tended to swim against the current. The moving stripes are probably generating the same types of reflex as the water current for the zebrafish. We could also see that the mean speed and inter-individual distances were slightly increased in the biased setup, which is often the case for a shoal of fish in a water current (Suriyampola et al. 2016).

We could show that, in a non-biased circular arena, we succeeded in building a mixed society, composed of half animal agents and half robots, with the robots able to modulate the choices of the zebrafish. When the robots were programmed to follow the collective decision of the fish, the mixed society of three robots and three fish behaved the same way as a group of six zebrafish that were free to choose their swimming direction. In the case where the robots were programmed to only swim in one direction, they influenced the collective decisions of the fish, and thus the whole mixed society collective choice was changed significantly.

We finally showed that, in a setup biased by the rotative stripes machine, we could also modulate the collective decisions of the fish using the robotic agents. However, due to the bias induced by the rotative stripes machine, the modulation was not the same compared to that obtained in the non-biased setup. However, this shows that the zebrafish were more cohesive with the decisions of the robots when the robots were programmed to follow their decisions instead of imposing a decision. This indicates that, in order to integrate robots into a group of zebrafish, the robots need to adapt and swim with the fish in order to increase their ability to modulate their decisions. This is a strong motivation for continuing efforts aimed to build accurate controllers for the robots to mimic the behavior of fish and swim with them in the shoal.

The measures of the fish linear speed and inter-individual distances confirmed that the robots were not inducing stress in the fish. However, the behavior of the fish still changed slightly, and thus improvements could still be made in terms of the lure design and robot control to achieve a complete integration of the robots into the group.

Among the different groups that are working in the field of fish-robot interaction, we are the first being able to develop a mixed group composed of more than two robotics agents. We are also the first to have designed a closed-loop system with mixed groups composed of half robots and half fish, in which the robots can be controlled according to the current fish behavior while imposing a behavior on the fish at the same time. It appears that the robots are also integrated into the mixed groups. Finally, we have shown a comparison between a non-biased and biased setup, highlighting the potential of using robotic agents inserted into animal societies to sense environmental changes through changes in animal behavior.

To summarize, the results shown in this chapter are very promising for the field of fish-robot interaction. Despite the fact that the design of the experiments involved a constrained environment for the fish in order to retrieve a clear collective response, we hope that these results will be extended in further studies in more complex environments, involving societies composed of more agents, in order to study how information can be transferred between the agents, how the robots can accurately modulate the behaviors of the fish, and possibly, how the robots can adapt to these behaviors in more detail.

11.5 People Who Contributed to This Work

Matthieu Broisin contributed to the design of the rotative stipes machine and performed the first tests to qualify the system for the experiments with zebrafish.

Chapter 12
Conclusion

12.1 Summary

In this chapter, we conclude on the global achievements that were obtained throughout this study. We summarize the main contributions of this thesis in the topic of animal-robot interaction studies and present the perspective for the future in this field. We also mention the lessons learned about the use of zebrafish for this type of study and the potential applications of our designs and discoveries.

12.2 Main Accomplishments

This work provided several main contributions to the state of the art. On a global level, we have contributed to the field of animal-robot interaction by showing a novel implementation of a mixed society of animals and robots based on the methodology proposed by Mondada et al. (2012), using societies composed of multiple vertebrates and robots.

In more detail:

- The first contribution concerns the translation of biological definitions into a modular robotic system, composed of a miniature wheeled mobile robot and an actuated biomimetic fish lure (Fig. 12.1). Due to its small size, reactive motions and continuous powering, the wheeled robot, FishBot, offers a convenient tool for biologists to study the effect of any type of lure inside a tank for experiments involving small fish species. The biomimetic actuated robotic lure, RiBot, is by far the smallest designed robotic lure currently used for such studies. As it is capable of imitating the movements of the tail, it offers a tool for biologists to measure the effect of the tail movements on the acceptance of a robot inside a fish society. We also proposed innovative design techniques for the conception of the RiBot, based on extracted 3D scans of real zebrafish and by designing the electronics using Rigid-Flex PCBs

F. Bonnet and F. Mondada, *Shoaling with Fish: Using Miniature Robotic Agents to Close the Interaction Loop with Groups of Zebrafish Danio rerio*, Springer Tracts in Advanced Robotics 131, https://doi.org/10.1007/978-3-030-16781-3_12

Fig. 12.1 The FishBot coupled with the RiBot during an exhibition at the Ars Electronica festival 2016 in Linz, Austria

in order to integrate more components while decreasing the size of the device. This robotic platform is modular in a way that the FishBot can be coupled with different types of lures designed to interact with different small species of fish including the RiBot. The RiBot can also be used alone if one wants to study only the stimuli generated by this device. The entire framework developed during this project is completely open-source and open-hardware.

- The second contribution is the design of a fully automated setup to perform multi-agent experiments involving small species of fish and robots, especially concerning the software CATS, which offers convenient control of multiple robots and tracking of animals for behavioral studies. Due mainly to its modularity, this software could be used in many fields of research for which closed-loop control of robots based on visual tracking is needed.
- The third contribution is the development of a bioinspired controller to reproduce the locomotion patterns of zebrafish based on data collected from the animal. We show that groups of robotic devices, when coupled with a model that simulates the typical trajectories of zebrafish inside the shoal, could mimic the group of fish in terms of trajectories and locomotion inside an open arena.
- The fourth contribution is our study of the acceptance of different types of lures among a shoal of zebrafish, when varying different parameters. Thanks to the fractional factorial design method, we could test the effect of various factors without having to test all the combinations, and thus use a lower quantity of animals. We showed that among different parameters, the lure motion as well as the lure shape had a strong effect on the acceptance of the lure in the fish shoal. By isolating the fish shape and tail-beating movements using a constrained setup composed of a circular corridor with a coaxial motor, we showed how the aspect of the lure as well as the tail beating could also increase the lure acceptance. The combination of biomimetic lure controlled using a biomimetic displacement model could thus be sufficient to integrate artificial agents inside fish shoals.

- Finally, our fifth contribution is the design and validation of an experiment involving a mixed society of fish and robots. This is the first time that a group of robots has infiltrated a group of vertebrates and had the ability to change the collective behavior of the animals. This mixed society of fish and robots is the result of the work performed on the development of the robotic agents and the analysis of the stimuli that were perceived by the fish under study. We were the first to design an experiment composed of more than two robots driven by bio-inspired controllers swimming among fish, which could at the same time demonstrate the integration of the robots among the animal society and with robots able to influence the collective decisions of the fish.

12.3 Lessons Learned in Robotic Design for Animal-Robot Interaction Studies

During this project, we designed miniature waterproof robots embedded with many different components such as stepper motors, LEDs and infrared receiver. We developed techniques to make the devices rechargeable and easy to build. We showed that very simple techniques based mainly on the molding of different materials could allow the miniaturization of the device while guaranteeing its waterproofness. Even though we were not able during the time of this thesis to design a fully autonomous robotic fish lure that can reproduce the movements of the zebrafish, we think that the device created is sufficient to perform many different types of experiments involving mixed societies of fish and robots. We also believe that in a near future, with the miniaturization of electronic et mechanical components, it will be possible to create an autonomous robotic lure with the same size of a zebrafish, that can at the same time reproduce the fish body movements and motion underwater.

We observed how the control of the robots can become difficult when introducing multiple robots, with the robots' movements also constrained by the small size of the environment. We showed how the use of simple navigation algorithms for mobile robots could resolve these issues while maintaining biomimetic locomotion. Biomimetic controllers that were designed preliminary without taking the robot dynamic into account could be applied to our robotic system, mainly because the specifications were well drawn for the robot design at the beginning of the project.

During the robots' development phases, we realized several times how the designed system was sometimes complex to use, as we encountered problems such as the FishBots decoupling from the lures during an experiment, losing power after several hours of experiments, and disconnecting from Bluetooth. One of the major issues that we encountered during the first years of the project was also due to the distance between Paris, where most of the experiments took place, and Lausanne, where the technical infrastructure was built. A lot of time and resources were spent on long-distance discussions and traveling. We think that to improve the project's efficiency, the collaborators should be located in closer proximity to each other, as the robotic design should be realized in very tight collaboration with the biologists, with both parties meeting on a regular basis.

In terms of the collaboration between engineers and biologists, we could notice how a multidisciplinary project with people from different fields of expertise could benefit for the global achievements obtained. Indeed, biologists were particularly demanding in terms of the specifications of the robotic system, as for them it should mimic as closely as possible all aspects of the animal, and also be capable of generating results that are repeatable. Engineers, on the other hand, could find the trade-off between efficiency and simplicity, for instance by testing new types of arenas and new configurations of the robotic systems and suggesting them to the biologists. Thus, we think that such collaborations are essential for this field of study and could only benefit all partners.

12.4 Lessons Learned from Zebrafish Social Behavior

Using our robotic design, we showed that, in order to be accepted inside the fish shoal and able to monitor collective decisions of the zebrafish, each robotic agent should have several important aspects:

- It should be able to swim among the fish shoal, either by using models that the robot can follow or by constraining the setup.
- It should have a biomimetic appearance. However, it is still difficult for us to claim which aspects have the most impact. For instance, it has been shown that for the guppies, the eye is an important biomimetic feature that can increase lure acceptance (Landgraf et al. 2016). In the case of the zebrafish, some studies suggested that certain aspects such as the stripes and the size ratio could increase lure acceptance (Abaid et al. 2012a). However, we show that a lure without stripes could be attractive enough to significantly modify the fish collective behavior for long periods of time.
- Moving parts on the body of the lure could increase its acceptance, as the tests involving the RiBot could demonstrate. Compared with what was shown in Abaid et al. (2012a), the robotic fish lure RiBot could reproduce tail beating frequencies more closely to those produced by the zebrafish, showing that such frequencies were also acceptable. We also shown that the fishing lure equipped with a passive tail beating was also the most attractive among all the passive lures that were tested.
- It should have a locomotion pattern mimicking the fish locomotion. For instance, in a corridor, it should move at a constant speed and follow the walls, and in an open arena, it should move with jerks using the same movement phase of zebrafish: orientation, acceleration and deceleration.

The system that was designed in this work can provide all these points, and we showed in our last experiments that, indeed, a shoal of three robots seemed to be infiltrated among the fish shoal and had the ability to modify the fish collective decisions.

However, different aspects can still be improved. For instance by controlling the robots using more biomimetic model to mimic the shoaling behavior of the fish. Also, we did not test the impact of the height of the lure, as the lure is moving at a fixed height. The difficulty is in knowing what aspects are playing a role and at which level, but this is more a question for the biologists who will use this framework in the future and will investigate more deeply the effect of all the different factors, requiring years of experiments. For instance, by making experiments longer, using different strains of zebrafish, these parameters could be compared. In a roboticist point of view, we conclude that our devices were able to satisfy the requirements set by the biologists, and were validated to perform experiments with the animals.

12.5 The Fish, an Appropriate Subject for Behavioral Studies?

Recent studies have shown that the fish might be a good candidate for studying the collective behavior of animals, but compared, for instance, with social insects that clearly show what is called *Swarm Intelligence*, the social mechanisms and information transfer in fish might be more complex to study (Ioannou 2016).

There are also many fish behavioral studies in the literature showing a lack of robustness in our opinion. The recent studies performed by the group of Porfiri that tested fish reaction to lures of different sizes (Bartolini et al. 2016) showed many contradictory results compared to their previous studies, meaning that the variability of the fish behavior needs to be taken into account and the experiments carefully conducted. In this study, we also encountered a lot of variability in the results between experiments, mostly using experimental setups like the open arena (Fig. 6.5a), in which the behavior of the zebrafish may vary depending on the zone, as shown in Chap. 8. Therefore, we decided to perform our experiments with a higher number of repeated runs and lasting at least 30 min in order to obtain a more robust and repeatable measurement of zebrafish behavior. In addition, we are very confident that with our approach using the circular corridor we could analyze in a robust way the fish collective decisions. We had very robust and repeatable results with this setup, even when testing different groups of zebrafish and having different runs of experiments spread over several months. We believe that such a setup is a good candidate to test mixed groups of robots and animals due to its homogeneity.

In summary, we do believe that the zebrafish has real potential for behavioral studies using it as a model of social vertebrates, and for experiments involving mixed societies of fish and robots as the ones presented in this dissertation.

12.6 Potential Applications

The work presented in this dissertation involving animals was performed only in a laboratory using laboratory animals. The results are thus mostly constrained by the environment, and we will not generalize our work to all fish evolving in their natural environment. Moreover, we used a robotic system that can only evolve in a very specific setup. However, some results of this thesis could potentially be used in a different area of study.

12.6.1 *Deeper Behavioral Research on Fish*

Deeper studies on fish social aggregation in their natural environment could be beneficial at the same time for fishermen and for the fish themselves. Indeed, nowadays, people are using systems like Fish Aggregation Devices (FAD) to attract the fish into specific spots (Dempster and Taquet 2004). But, unfortunately, due to the fact that the reaction of the fish towards these systems is not well studied, many fish usually are captured even though the fishermen do not want them (Capello et al. 2011). Thus, a deeper understanding of fish behavior and fish reaction to specific stimuli is needed, and tools as the ones developed in this research could contribute to extending the knowledge of fish behavior.

In this research, we focused on the social behavior of the fish, but there are also other behavior that could be tested with the same apparatus. For instance, we could test the escaping behavior of the individuals using a predator fish lure as it is done in Swain et al. (2012), or a prey lure to analyze the predator behavior of fish. The fact that our robotic platform is modular allow to test different types of lures that can be controlled in a closed-loop.

Another application of behavioral study that came out into a concrete product is the repellent bracelet for sharks (Sha 2016). This bracelet uses a repulsive stimulus to repel the shark away from people. This is a good example of a non-invasive method of modifying the behavior of animals for the well-being of humans, and the types of study to extract signals that are sensitive for fish could be performed using the framework that we designed.

12.6.2 *Ecology*

We showed in the previous chapter that we could measure a change in the animal behavior when an environmental factor was included. This experiment showed that the animal sensing capabilities can be used to retrieve information from the environment and, moreover, that our robotic system could be used to modulate the behavior of fish depending on the environmental conditions. Fish today are suffering from pollutants and other negative environmental factors. By being able to sense animal behavior changes and modify them depending on the situation, we could simultaneously benefit from animal knowledge of the environment and possibly save the animals in cases in which they cannot find a solution on their own.

12.6.3 Pharmaceutical

Our system could be used to help identifying behavioral genes and neural connection responsible for behavior in vertebrates. For instance, in Stewart et al. (2014), they suggest the zebrafish as a model to study autism spectrum disorder. For such experiments, our robotic system could be used to identify behavior changes of zebrafish when modified genetically or using chemicals to cure the disease. Many other behaviors, such as aggression, anxiety and sleep, which are currently using zebrafish as a model as shown in Norton and Bally-Cuif (2010) could also use our robotic system to monitor and control fish behavior in very long-duration experiments.

12.6.4 Research Involving Collective Behavior Studies

Other groups are working with a similar circular corridor arena to test the collective behavior of agents. For instance, in Lemercier et al. (2012), they use it to study the effect of the crowd on human pedestrians. Crowds and traffic flow management are a challenge in our society when a mass of people aggregates towards the same points, and the modelling of crowd behavior using animals such as fish in a constrained environment could be extended to other social species, using optimization methods such as the ones described in Cazenille et al. (2015).

12.6.5 Benefits for the ASSISIbf Project

The robots, software and automated setup that were designed during this thesis were produced in order to be used by our main partner in the ASSISIbf project, the team of the University Paris Diderot that was involved in the experiments to study the collective behavior of zebrafish and to set up mixed societies of zebrafish and robots. In total, 30 FishBots and 30 RiBots were produced during the project and were or will be used in the ASSISIbf project, and hopefully after the project. Two automated setups were implemented in Paris on the same basis as the one implemented at the EPFL to study fish behavior.

The extension of the software with long-distance infrastructure (Sect. 7.10) allows the connection between the fish experimental setup in Paris and the bee experimental setup in Graz. This experiment, which is the final target experiment of ASSISIbf,

Fig. 12.2 Fish and bee mixed society setup at the ARS Electronica festival 2016 in Linz, Austria. On the left (black): the bee setup with the running experiment involving honeybees and robotic honeybees shown with a projector. On the right (white): the fish setup with the running experiment involving zebrafish and the robotic fish that was also projected on a screen

will be able to show how two social animal species, i.e. zebrafish and bees, can communicate through robotic agents, and how this multi-artificial-living system could evolve by learning from each other.

A public demonstration of the framework developed during this project was shown at the ARS Electronica Festival in Linz, Austria (Fig. 12.2). Live experiments showing mixed societies of fish and robots and mixed societies of bees and robots could be shown to the public. Experiments involving interactions between the two mixed societies were also shown during this festival (Fig. 12.3).

12.7 Final Words

To conclude, this work contributed to the animal-robot interaction research, more specifically the fish-robot interaction research, by miniaturizing the standard robotic tools used for these types of studies and adding multiple robots into the interaction loop, showing that these tools can be used to design mixed societies of robots and vertebrates. In addition, we showed that our hypothesis about the increase of lure acceptance on the fish using biomimetic cues, such as tail beating, locomotion patterns, fish motion model reproduction and visual aspects could be verified. Our goal was to create a mixed society of robots and fish, thus showing a new application of the general methodology that is used for such types of research, and the results presented in this dissertation showed that a group composed of half robots and half

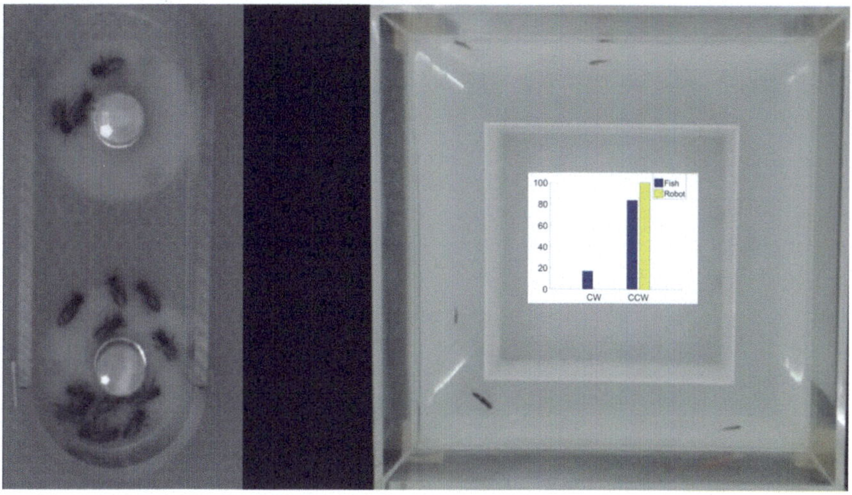

Fig. 12.3 Fish and bees interacting at the ARS Electronica festival 2016 in Linz, Austria. On the left: the bee arena with the honeybees that aggregate near one of the two bee robots. On the right: the fish arena with the fish following the lure that is steered by a FishBot, which is driven by the bees' collective decision

fish could be created, with the robots able to behave as zebrafish and also to change the collective behavior of the animals. Finally, by being introduced into the animal societies, the robots might be able to learn the language and behaviors of animals, and we hope that the tools presented in this study will be used to learn more about fundamental biological aspects of the collective behaviors in fish, taking interest in the shoaling behavior of fish to new levels.

References

(2013). Robofish, the fastest selling toy in 2013. http://www.dailymail.co.uk/femail/article-2513912/Robo-Fish-fastest-selling-toy-year-globally.html.

(2015). Why ROS 2.0? http://design.ros2.org/articles/why_ros2.html.

(2016). Aravis - a vision library for GenICam based cameras. https://github.com/AravisProject/aravis.

(2016). Ars electronica festival web page. http://www.aec.at/festival/.

(2016). ASEBA library. https://github.com/aseba-community/aseba.

(2016). Boost c++ library. http://www.boost.org/.

(2016). A fast single-producer, single-consumer lock-free queue for c++. https://github.com/cameron314/readerwriterqueue.

(2016). GStreamer - open source multimedia framework. https://gstreamer.freedesktop.org.

(2016). Molole, the mobots low level library. https://gna.org/projects/molole/.

(2016). Qt. cross-platform software development for embedded and desktop. https://www.qt.io.

(2016). QtDBus library to interface with ASEBA medulla. https://github.com/aseba-community/aseba/tree/master/examples/clients/cpp-qt-dbus.

(2016). Robotswim, Jessiko project. http://www.robotswim.com/index.php?lan=en.

(2016). Sharkbanz anti-shark bracelet. http://www.sharkbanz.com/home/.

(2016). Sources of the control and tracking software cats. https://github.com/gribovskiy/CATS2/.

(2017). Linux infrared remote control library (LIRC). http://www.lirc.org/.

Abaid, N., & Porfiri, M. (2010). Fish in a ring: Spatio-temporal pattern formation in one-dimensional animal groups. *Journal of the Royal Society Interface* (p. rsif20100175).

Abaid, N., Bartolini, T., Macri, S., & Porfiri, M. (2012a). Zebrafish responds differentially to a robotic fish of varying aspect ratio, tail beat frequency, noise, and color. *Behavioural Brain Research*, *233*(2), 545–553.

Abaid, N., Spinello, C., Laut, J., & Porfiri, M. (2012b). Zebrafish (Danio rerio) responds to images animated by mathematical models of animal grouping. *Behavioural Brain Research*, *232*(2), 406–410.

Abaid, N., Marras, S., Fitzgibbons, C., & Porfiri, M. (2013). Modulation of risk-taking behaviour in golden shiners (Notemigonus crysoleucas) using robotic fish. *Behavioural Processes*, *100*, 9–12.

Ashraf, I., Godoy-Diana, R., Halloy, J., Collignon, B., & Thiria, B. (2016). Synchronization and collective swimming patterns in fish (Hemigrammus bleheri). *Journal of the Royal Society Interface*, *13*(123).

Aureli, M., Fiorilli, F., & Porfiri, M. (2012). Portraits of self-organization in fish schools interacting with robots. *Physica D: Nonlinear Phenomena*, *241*(9), 908–920.

F. Bonnet and F. Mondada, *Shoaling with Fish: Using Miniature Robotic Agents to Close the Interaction Loop with Groups of Zebrafish Danio rerio*, Springer Tracts in Advanced Robotics 131, https://doi.org/10.1007/978-3-030-16781-3

Bandyopadhyay, P. R. (2005). Trends in biorobotic autonomous undersea vehicles. *IEEE Journal of Oceanic Engineering, 30*(1), 109–139.
Bartolini, T., Mwaffo, V., Showler, A., Macrì, S., Butail, S., & Porfiri, M. (2016). Zebrafish response to 3d printed shoals of conspecifics: The effect of body size. *Bioinspiration and Biomimetics, 11*(2), 026003.
Bass, S. L., & Gerlai, R. (2008). Zebrafish (Danio rerio) responds differentially to stimulus fish: The effects of sympatric and allopatric predators and harmless fish. *Behavioural Brain Research, 186*(1), 107–117.
Bonabeau, E., Dorigo, M., & Theraulaz, G. (1999). *Swarm intelligence: From natural to artificial systems* (Vol. 1). Oxford: Oxford University Press.
Bonnet, F., Rétornaz, P., Halloy, J., Gribovskiy, A., & Mondada, F. (2012). Development of a mobile robot to study the collective behavior of zebrafish. In *2012 4th IEEE RAS EMBS international conference on biomedical robotics and biomechatronics (BioRob)* (pp. 437–442).
Bonnet, F., Binder, S., de Oliveria, M. E., Halloy, J., & Mondada, F. (2014). A miniature mobile robot developed to be socially integrated with species of small fish. In *2014 IEEE international conference on robotics and biomimetics (ROBIO)* (pp. 747–752).
Bonnet, F., Crot, N., Burnier, D., & Mondada, F. (2016a). Design methods for miniature underwater soft robots. In *2016 6th IEEE international conference on biomedical robotics and biomechatronics (BioRob)* (pp. 1365–1370).
Bonnet, F., Kato, Y., Halloy, J., & Mondada, F. (2016b). Infiltrating the zebrafish swarm: Design, implementation and experimental tests of a miniature robotic fish lure for fish-robot interaction studies. *Artificial Life and Robotics, 21*(3), 239–246.
Bonnet, F., Cazenille, L., Gribovskiy, A., Halloy, J., & Mondada, F. (2017a). Multi-robots control and tracking framework for bio-hybrid systems with closed-loop interaction. In *IEEE international conference on robotics and automation*, submitted.
Bonnet, F., Cazenille, L., Séguret, A., Collignon, B., Gribovskiy, A., Halloy, J., et al. (2017b). Design of a modular robotic system that mimics small fish locomotion and body movements for ethological studies. *International Journal of Advanced Robotic Systems*, accepted.
Bradski, G. (2000). Dr. Dobb's Journal of Software Tools.
Braitenberg, V. (1986). *Vehicles: Experiments in synthetic psychology*. Cambridge: MIT Press.
Butail, S., Bartolini, T., & Porfiri, M. (2013). Collective response of zebrafish shoals to a free-swimming robotic fish. *PLoS One, 8*(10), e76123.
Butail, S., Ladu, F., Spinello, D., & Porfiri, M. (2014a). Information flow in animal-robot interactions. *Entropy, 16*(3), 1315–1330.
Butail, S., Polverino, G., Phamduy, P., Del Sette, F., & Porfiri, M. (2014b). Fish-robot interactions in a free-swimming environment: Effects of speed and configuration of robots on live fish, *9055*, 90550I–90550I–7.
Camazine, S., Deneubourg, J.-L., Franks, N. R., Sneyd, J., Theraulaz, G., & Bonabeau, E. (2001). *Self-organization in biological systems*. Princeton: Princeton University Press.
Capello, M., Soria, M., Cotel, P., Deneubourg, J.-L., & Dagorn, L. (2011). Quantifying the interplay between environmental and social effects on aggregated-fish dynamics. *PloS One, 6*(12), e28109.
Cazenille, L., Bredeche, N., & Halloy, J. (2015). *Multi-objective optimization of multi-level models for controlling animal collective behavior with robots* (pp. 379–390). Cham: Springer International Publishing.
Chablais, F., Veit, J., Rainer, G., & Jaźwińska, A. (2011). The zebrafish heart regenerates after cryoinjury-induced myocardial infarction. *BMC Developmental Biology, 11*(1), 1.
Chen, Z., Shatara, S., & Tan, X. (2010). Modeling of biomimetic robotic fish propelled by an ionic polymer-metal composite caudal fin. *IEEE/ASME Transactions on Mechatronics, 15*(3), 448–459.
Collignon, B., Séguret, A., & Halloy, J. (2016). A stochastic vision-based model inspired by zebrafish collective behaviour in heterogeneous environments. *Royal Society Open Science, 3*(1), 150473.
Corp, M. S. (2016). Merlin robotics. http://www.merlinsystemscorp.co.uk/.

Correll, N., Schwager, M., & Rus, D. (2008). Social control of herd animals by integration of artificially controlled congeners. In M. Asada, J. C. T. Hallam, J.-A. Meyer, & J. Tani (Eds.), *SAB*. Lecture notes in computer science (Vol. 5040, pp. 437–446). Berlin: Springer.

Crowe, C. T., Elger, D. F., Williams, B. C., & Roberson, J. A. (2010). *Engineering fluid mechanics*. New York: Wiley.

Czaczkes, T. J., & Ratnieks, F. L. (2013). Cooperative transport in ants (Hymenoptera: Formicidae) and elsewhere. *Myrmecological News*, *18*, 1–11.

Delcourt, J., Denoël, M., Ylieff, M., & Poncin, P. (2013). Video multitracking of fish behaviour: A synthesis and future perspectives. *Fish and Fisheries*, *14*(2), 186–204.

Dempster, T., & Taquet, M. (2004). Fish aggregation device (FAD) research: Gaps in current knowledge and future directions for ecological studies. *Reviews in Fish Biology and Fisheries*, *14*(1), 21–42.

Detrain, C., & Deneubourg, J.-L. (2006). Self-organized structures in a superorganism: Do ants "behave" like molecules? *Physics of Life Reviews*, *3*(3), 162–187.

Dijkstra, E. W. (1959). A note on two problems in connexion with graphs. *Numerische Mathematik*, *1*(1), 269–271.

Donati, E., Worm, M., Mintchev, S., van der Wiel, M., Benelli, G., von der Emde, G., et al. (2016). Investigation of collective behaviour and electrocommunication in the weakly electric fish, Mormyrus rume, through a biomimetic robotic dummy fish. *Bioinspiration and Biomimetics*, *11*(6), 066009.

Dorigo, M., Birattari, M., & Stutzle, T. (2006). Ant colony optimization. *IEEE Computational Intelligence Magazine*, *1*(4), 28–39.

Du, R., Li, Z., Youcef-Toumi, K., & Valdivia y Alvarado, P. (2015). *Robot fish: Bio-inspired fishlike underwater robots*. Berlin: Springer.

Dykgraaf, S. (1933). Untersuchungen über die funktion der seitenorgane an fischen. *Zeitschrift für vergleichende Physiologie*, *20*(1), 162–214.

Faria, J. J., Dyer, J. R., Clément, R. O., Couzin, I. D., Holt, N., Ward, A. J., et al. (2010). A novel method for investigating the collective behaviour of fish: Introducing 'robofish'. *Behavioral Ecology and Sociobiology*, *64*(8), 1211–1218.

Fleisch, V. C., & Neuhauss, S. C. (2006). Visual behavior in zebrafish. *Zebrafish*, *3*(2), 191–201.

Furbringer, J.-M. (2005). Design of experiments. Doctoral courses.

Garnier, S. (2011). *From ants to robots and back: How robotics can contribute to the study of collective animal behavior* (pp. 105–120). Berlin: Springer.

Gemberling, M., Bailey, T. J., Hyde, D. R., & Poss, K. D. (2013). The zebrafish as a model for complex tissue regeneration. *Trends in Genetics*, *29*(11), 611–620.

Giattina, J., & Garton, R. (1983). A review of the preference-avoidance responses of fishes to aquatic contaminants. *Residue reviews* (pp. 43–90). Berlin: Springer.

Goulson, D., Nicholls, E., Botías, C., & Rotheray, E. L. (2015). Bee declines driven by combined stress from parasites, pesticides, and lack of flowers. *Science*, *347*(6229), 1255957.

Gribovskiy, A., & Mondada, F. (2010). Animal-robot interaction for ethological studies: An advance framework based on socially integrated mobile robots

Gribovskiy, A., Halloy, J., Deneubourg, J.-L., & Mondada, F. (2010). The PoulBot: A mobile robot for ethological studies on domestic chickens. In *Symposium on AI-inspired biology (AIIB)*.

Griparic, K., Haus, T., Miklic, D., & Bogdan, S. (2015). Combined actuator sensor unit for interaction with honeybees. In *2015 IEEE sensors applications symposium (SAS)* (pp. 1–5).

Halloy, J., Sempo, G., Caprari, G., Rivault, C., Asadpour, M., Tache, F., et al. (2007). Social integration of robots into groups of cockroaches to control self-organized choices. *Science*, *318*(5853), 1155–1158.

Halloy, J., Mondada, F., Kernbach, S., & Schmickl, T. (2013). *Towards bio-hybrid systems made of social animals and robots* (pp. 384–386). Berlin: Springer.

Heo, S., Wiguna, T., Park, H. C., & Goo, N. S. (2007). Effect of an artificial caudal fin on the performance of a biomimetic fish robot propelled by piezoelectric actuators. *Journal of Bionic Engineering*, *4*(3), 151–158.

Hoare, D. J., Couzin, I. D., Godin, J. G. J., & Krause, J. (2004). Context-dependent group size choice in fish. *Animal Behaviour*, *67*(1), 155–164.

Howe, K., Clark, M. D., Torroja, C. F., Torrance, J., Berthelot, C., Muffato, M., et al. (2013). The zebrafish reference genome sequence and its relationship to the human genome. *Nature*, *496*(7446), 498–503.

Hunter, W. G., Hunter, J. S., et al. (1978). *Statistics for experimenters: An introduction to design, data analysis, and model building*. New York: Wiley.

Ioannou, C. C. (2016). Swarm intelligence in fish? The difficulty in demonstrating distributed and self-organised collective intelligence in (some) animal groups. *Behavioural Processes*.

KaewTraKulPong, P., & Bowden, R. (2002). An improved adaptive background mixture model for real-time tracking with shadow detection. *Video-based surveillance systems: Computer vision and distributed processing* (pp. 135–144). Berlin: Springer.

Kalman, R. E. (1960). A new approach to linear filtering and prediction problems. *Journal of Basic Engineering*, *82*(1), 35–45.

Kawabata, K., Aonuma, H., Hosoda, K., & Xue, J. (2013). A system for automated interaction with the cricket utilizing a micro mobile robot. *Journal of Robotics and Mechatronics*, *25*(2), 333–339.

Khatib, O. (1986). Real-time obstacle avoidance for manipulators and mobile robots. *The International Journal of Robotics Research*, *5*(1), 90–98.

Kleinhappel, T. K., Burman, O. H., John, E. A., Wilkinson, A., & Pike, T. W. (2016). A mechanism mediating inter-individual associations in mixed-species groups. *Behavioral Ecology and Sociobiology*, *70*(5), 755–760.

Kopman, V., & Porfiri, M. (2013). Design, modeling, and characterization of a miniature robotic fish for research and education in biomimetics and bioinspiration. *IEEE/ASME Transactions on Mechatronics*, *18*(2), 471–483.

Krause, J., Hoare, D., Krause, S., Hemelrijk, C., & Rubenstein, D. (2000). Leadership in fish shoals. *Fish and Fisheries*, *1*(1), 82–89.

Krause, J., Winfield, A. F., & Deneubourg, J.-L. (2011). Interactive robots in experimental biology. *Trends in Ecology and Evolution*, *26*(7), 369–375.

Ladu, F., Bartolini, T., Panitz, S., Chiarotti, F., Butail, S., Macrì, S., et al. (2015a). Live predators, robots, and computer-animated images elicit differential avoidance responses in zebrafish. *Zebrafish*.

Ladu, F., Mwaffo, V., Li, J., Macrì, S., & Porfiri, M. (2015b). Acute caffeine administration affects zebrafish response to a robotic stimulus. *Behavioural Brain Research*, *289*, 48–54.

Landgraf, T., Oertel, M., Rhiel, D., & Rojas, R. (2010). A biomimetic honeybee robot for the analysis of the honeybee dance communication system. In *Proceedings of the IEEE/RSJ 2010 international conference on intelligent robots and systems (IROS 2010)*.

Landgraf, T., Nguyen, H., Forgo, S., Schneider, J., Schröer, J., Krüger, C., et al. (2013). Interactive robotic fish for the analysis of swarm behavior. *Advances in swarm intelligence: 4th international conference, ICSI 2013, Harbin, China, 12–15 June 2013, proceedings, part I*. Berlin: Springer.

Landgraf, T., Bierbach, D., Nguyen, H., Muggelberg, N., Romanczuk, P., & Krause, J. (2016). Robofish: Increased acceptance of interactive robotic fish with realistic eyes and natural motion patterns by live Trinidadian guppies. *Bioinspiration and Biomimetics*, *11*(1), 015001.

Le Maho, Y., Whittington, J. D., Hanuise, N., Pereira, L., Boureau, M., Brucker, M., et al. (2014). Rovers minimize human disturbance in research on wild animals. *Nature Methods*, *11*(12), 1242–1244.

Lemasson, B. H., Anderson, J. J., & Goodwin, R. A. (2013). Motion-guided attention promotes adaptive communications during social navigation. *Proceedings of the Royal Society of London B: Biological Sciences*, *280*(1757).

Lemercier, S., Jelic, A., Kulpa, R., Hua, J., Fehrenbach, J., Degond, P., et al. (2012). Realistic following behaviors for crowd simulation. *Computer Graphics Forum*, *31*, 489–498. Wiley Online Library.

Maaswinkel, H., & Li, L. (2003). Spatio-temporal frequency characteristics of the optomotor response in zebrafish. *Vision Research*, *43*(1), 21–30.

Madakam, S., Ramaswamy, R., & Tripathi, S. (2015). Internet of things (IoT): A literature review. *Journal of Computer and Communications*, *3*(05), 164.

Magalhães, D. d. P., Da Cunha, R. A., dos Santos, J. A. A., Buss, D. F., & Baptista, D. F. (2007). Behavioral response of zebrafish Danio rerio Hamilton 1822 to sublethal stress by sodium hypochlorite: Ecotoxicological assay using an image analysis biomonitoring system. *Ecotoxicology*, *16*(5), 417–422.

Magalhães, D. d. P., Buss, D. F., Da Cunha, R. A., Linde-Arias, A. R., & Baptista, D. F. (2012). Analysis of individual versus group behavior of zebrafish: A model using pH sublethal effects. *Bulletin of Environmental Contamination and Toxicology*, *88*(6), 1009–1013.

Magnenat, S., & Mondada, F. (2009). ASEBA meets D-bus: From the depths of a low-level event. In *IEEE TC-soft workshop on event-based systems in robotics (EBS-RO)*. Invited paper.

Magnenat, S., Retornaz, P., Bonani, M., Longchamp, V., & Mondada, F. (2011). ASEBA: A modular architecture for event-based control of complex robots. *IEEE/ASME Transactions on Mechatronics*, *16*(2), 321–329.

Marchese, A. D., Onal, C. D., & Rus, D. (2013). Towards a self-contained soft robotic fish: On-board pressure generation and embedded electro-permanent magnet valves. *Experimental robotics* (pp. 41–54). Berlin: Springer.

Marras, S., & Porfiri, M. (2012). Fish and robots swimming together: Attraction towards the robot demands biomimetic locomotion. *Journal of the Royal Society Interface*, *9*(73), 1856–1868.

Mondada, F., Halloy, J., Martinoli, A., Correll, N., Gribovskiy, A., Sempo, G., et al. (2012). A general methodology for the control of mixed natural-artificial societies. *Handbook of collective robotics*. To appear: Pan Stanford.

Mondada, F., Bonnet, E., Davrajh, S., Johal, W., & Stopforth, R. (2016). R2T2 robotics to integrate educational efforts in South Africa and Europe. *International Journal of Advanced Robotic Systems*, *13*(5), 1–13.

Montgomery, D. C. (2008). *Design and analysis of experiments*. New York: Wiley.

Munack, A. (2002). Agriculture and the environment: New challenges for engineers. *Agricultural Engineering International: CIGR Journal*.

Mwaffo, V., Anderson, R. P., Butail, S., & Porfiri, M. (2014). A jump persistent turning walker to model zebrafish locomotion. *Journal of the Royal Society Interface*, *12*(102).

Norton, W., & Bally-Cuif, L. (2010). Adult zebrafish as a model organism for behavioural genetics. *BMC Neuroscience*, *11*(1), 1.

Olson, R. S., Haley, P. B., Dyer, F. C., & Adami, C. (2015). Exploring the evolution of a trade-off between vigilance and foraging in group-living organisms. *Royal Society Open Science*, *2*(9), 150135.

Pérez-Escudero, A., Vicente-Page, J., Hinz, R. C., Arganda, S., & de Polavieja, G. G. (2014). idtracker: Tracking individuals in a group by automatic identification of unmarked animals. *Nature Methods*, *11*(7), 743–748.

Perry, S. F., Ekker, M., Farrell, A. P., & Brauner, C. J. (2010). *Fish physiology: Zebrafish* (Vol. 29). New York: Academic.

Phamduy, P., Polverino, G., Fuller, R., & Porfiri, M. (2014). Fish and robot dancing together: Bluefin killifish females respond differently to the courtship of a robot with varying color morphs. *Bioinspiration and Biomimetics*, *9*(3), 036021.

Pitcher, T. J., & Parrish, J. K. (1993). Functions of shoaling behaviour in teleosts. *The behaviour of teleost fishes* (pp. 294–337). London: Chapman and Hall.

Polverino, G., & Porfiri, M. (2013a). Mosquitofish (Gambusia affinis) responds differentially to a robotic fish of varying swimming depth and aspect ratio. *Behavioural Brain Research*, *250*, 133–138.

Polverino, G., & Porfiri, M. (2013b). Zebrafish (Danio rerio) behavioural response to bioinspired robotic fish and mosquitofish (Gambusia affinis). *Bioinspiration and Biomimetics*, *8*(4), 044001.

Polverino, G., Abaid, N., Kopman, V., Macrì, S., & Porfiri, M. (2012). Zebrafish response to robotic fish: Preference experiments on isolated individuals and small shoals. *Bioinspiration and Biomimetics*, *7*(3), 036019.

Poss, K. D., Wilson, L. G., & Keating, M. T. (2002). Heart regeneration in zebrafish. *Science, 298*(5601), 2188–2190.

Poss, K. D., Keating, M. T., & Nechiporuk, A. (2003). Tales of regeneration in zebrafish. *Developmental Dynamics, 226*(2), 202–210.

Quigley, M., Gerkey, B., Conley, K., Faust, J., Foote, T., Leibs, J., et al. (2009). ROS: An open-source robot operating system. In *Proceedings of the IEEE international conference on robotics and automation (ICRA) workshop on open source robotics*.

Reed, B., & Jennings, M. (2011). *Guidance on the housing and care of zebrafish*. Southwater: Royal Society for the Prevention of Cruelty to Animals.

Rosenthal, G. G., & Ryan, M. J. (2005). Assortative preferences for stripes in danios. *Animal Behaviour, 70*(5), 1063–1066.

Rosenthal, S. B., Twomey, C. R., Hartnett, A. T., Wu, H. S., & Couzin, I. D. (2015). Revealing the hidden networks of interaction in mobile animal groups allows prediction of complex behavioral contagion. *Proceedings of the National Academy of Sciences, 112*(15), 4690–4695.

Rundus, A. S., Owings, D. H., Joshi, S. S., Chinn, E., & Giannini, N. (2007). Ground squirrels use an infrared signal to deter rattlesnake predation. *Proceedings of the National Academy of Sciences, 104*(36), 14372–14376.

Sato, H., & Maharbiz, M. (2010). Recent developments in the remote radio control of insect flight. *Frontiers in Neuroscience, 4*, 199.

Saverino, C., & Gerlai, R. (2008). The social zebrafish: Behavioral responses to conspecific, heterospecific, and computer animated fish. *Behavioural Brain Research, 191*(1), 77–87.

Schmickl, T., Bogdan, S., Correia, L., Kernbach, S., Mondada, F., Bodi, M., et al. (2013). *ASSISI: Mixing animals with robots in a hybrid society* (pp. 441–443). Berlin: Springer.

Schroeder, P., Jones, S., Young, I. S., & Sneddon, L. U. (2014). What do zebrafish want? Impact of social grouping, dominance and gender on preference for enrichment. *Laboratory Animals* (p. 0023677214538239).

Séguret, A., Collignon, B., & Halloy, J. (2016). Strain differences in the collective behaviour of zebrafish (Danio rerio) in heterogeneous environment. *Royal Society Open Science, 3*(10).

Shi, J., & Tomasi, C. (1994). Good features to track. In *1994 IEEE computer society conference on computer vision and pattern recognition, 1994. Proceedings CVPR'94* (pp. 593–600). IEEE.

Shi, Q., Miyagishima, S., Konno, S., Fumino, S., Ishii, H., Takanishii, A. (2010). Development of the hybrid wheel-legged mobile robot WR-3 designed to interact with rats. In *3rd IEEE RAS and EMBS international conference on biomedical robotics and biomechatronics (BioRob)* (pp. 887 –892).

Shi, Q., Ishii, H., Sugahara, Y., Takanishi, A., Huang, Q., & Fukuda, T. (2015). Design and control of a biomimetic robotic rat for interaction with laboratory rats. *IEEE/ASME Transactions on Mechatronics, 20*(4), 1832–1842.

Spence, R., Gerlach, G., Lawrence, C., & Smith, C. (2008). The behaviour and ecology of the zebrafish, Danio rerio. *Biological Reviews of the Cambridge Philosophical Society, 83*(1), 13–34.

Spinello, C., Macrì, S., & Porfiri, M. (2013). Acute ethanol administration affects zebrafish preference for a biologically inspired robot. *Alcohol, 47*(5), 391–398.

Stewart, A. M., Nguyen, M., Wong, K., Poudel, M. K., & Kalueff, A. V. (2014). Developing zebrafish models of autism spectrum disorder (ASD). *Progress in Neuro-Psychopharmacology and Biological Psychiatry, 50*, 27–36.

Strandburg-Peshkin, A., Twomey, C. R., Bode, N. W., Kao, A. B., Katz, Y., Ioannou, C. C., et al. (2013). Visual sensory networks and effective information transfer in animal groups. *Current Biology, 23*(17), R709–R711.

Streisinger, G., Walker, C., Dower, N., Knauber, D., & Singer, F. (1981). Production of clones of homozygous diploid zebra fish (Brachydanio rerio). *Nature, 291*(5813), 293.

Suriyampola, P. S., Sykes, D. J., Khemka, A., Shelton, D. S., Bhat, A., & Martins, E. P. (2016). Water flow impacts group behavior in zebrafish (Danio rerio). *Behavioral Ecology*.

Swain, D. (2011). MadTrac framework source documentation.

Swain, D. T., Couzin, I. D., & Leonard, N. E. (2012). Real-time feedback-controlled robotic fish for behavioral experiments with fish schools. *Proceedings of the IEEE, 100*(1), 150–163.

Takada, Y., Araki, R., Nakanishi, Y., Nonogaki, M., Ebita, K., & Wakisaka, T. (2010). Development of small fish robots powered by small and ultra-light passive-type polymer electrolyte fuel cells. *Journal of Robotics and Mechatronics, 22*(2), 150.

Taylor, R. C., Klein, B. A., Stein, J., & Ryan, M. J. (2008). Faux frogs: Multimodal signalling and the value of robotics in animal behaviour. *Animal Behaviour, 76*(3), 1089–1097.

Tinbergen, N. (1951). *The study of instinct*. Oxford: Oxford University Press.

Travis, C. C. (2013). *Use of biomarkers in assessing health and environmental impacts of chemical pollutants* (Vol. 250). New York: Springer Science and Business Media.

Vaughan, R. T., Sumpter, N., Henderson, J., Frost, A., & Cameron, S. (2000). Experiments in automatic flock control. *Robotics and Autonomous Systems, 31*(1–2), 109–117.

Wang, T. (2017). Collision-free path planning using potential field method for highly redundant manipulators.

Wang, Z., Hang, G., Li, J., Wang, Y., & Xiao, K. (2008). A micro-robot fish with embedded sma wire actuated flexible biomimetic fin. *Sensors and Actuators A: Physical, 144*(2), 354–360.

Ward, A. J., Herbert-Read, J. E., Jordan, L. A., James, R., Krause, J., Ma, Q., et al. (2013). Initiators, leaders, and recruitment mechanisms in the collective movements of damselfish. *The American Naturalist, 181*(6), 748–760.

Wong, B. B. M., & Rosenthal, G. G. (2005). Shoal choice in swordtails when preferences conflict. *Ethology, 111*(2), 179–186.

Worm, M., Landgraf, T., Nguyen, H., & von der Emde, G. (2014). Electro-communicating dummy fish initiate group behavior in the weakly electric fish Mormyrus rume. *Conference on biomimetic and biohybrid systems* (pp. 446–448). Berlin: Springer.

Wright, D., Ward, A., Croft, D., & Krause, J. (2006). Social organization, grouping, and domestication in fish. *Zebrafish, 3*(2), 141–155.

Zeitoun, M. M., & Mehana, E.-S. E. (2014). Impact of water pollution with heavy metals on fish health: Overview and updates. *Global Veterinaria, 12*(2), 219–231.

Zienkiewicz, A., Barton, D. A., Porfiri, M., & di Bernardo, M. (2015). Data-driven stochastic modelling of zebrafish locomotion. *Journal of Mathematical Biology, 71*(5), 1081–1105.

The manufacturer's authorised representative in the EU is Springer Nature Customer Service Centre GmbH, Europaplatz 3, 69115 Heidelberg, Germany. If you have any concerns regarding our products, please contact ProductSafety@springernature.com

Printed and bound by CPI Group (UK) Ltd, Croydon, CR0 4YY
24/07/2026
02175907-0004